SpringerBriefs in Cell Biology

More information about this series at http://www.springer.com/series/10708

Qing Yan

Cellular Rhythms and Networks

Implications for Systems Medicine

 Springer

Qing Yan
PharmTao
Santa Clara, CA, USA

SpringerBriefs in Cell Biology
ISBN 978-3-319-22818-1   ISBN 978-3-319-22819-8   (eBook)
DOI 10.1007/978-3-319-22819-8

Library of Congress Control Number: 2015947931

Springer Cham Heidelberg New York Dordrecht London

Printed on acid-free paper

Springer International Publishing AG Switzerland is part of Springer Science+Business Media (www.springer.com)

Preface

Biological rhythms play pivotal roles in both physical and mental health. Spatiotemporal oscillations have been identified at different levels, from mitochondria to transmembrane potentials, from heart excitation waves to neural activities. The circadian clock is involved in gene expression regulations and various cellular processes including metabolism, proliferation, and senescence. This book provides an overview of the cellular rhythms and networks with the emphasis on the systems biology understanding of their roles in the practice of personalized and systems medicine.

The disruption of circadian rhythms has been associated with many complex diseases including insomnia, depression, heart disease, cancer, rheumatoid arthritis, and neurodegenerative disorders. The multi-scale view of circadian systems on the basis of systems biology would empower the discovery of novel therapeutic strategies such as chronotherapy (see Chap. 1). Depending on the feedback loops with multiple pathways and protein–protein interactions involved, the circadian clocks form the essential cellular timing mechanisms that synchronize vital physiological processes (see Chap. 2). For example, the multi-factorial circadian-neuroendocrine-immune networks may be involved in various disorders including the lung, heart, and gastrointestinal diseases (see Chap. 3).

Depression and circadian disruptions may share a common etiology with lower cellular resilience and reduced resistance to stressful events. The pattern analysis of systemic circadian profiles can be useful for the prediction and prevention of various psychiatric disorders (see Chap. 4). The cardiovascular system responds to environmental stimuli with circadian patterns. Such patterns are mediated via the complex interactions between the extracellular factors such as neuro-humoral elements and intracellular factors such as the clock genes with impacts on the pharmacokinetics and pharmacodynamics of drugs (see Chap. 5). Studies at molecular, cellular, and clinical levels have demonstrated the critical role of the circadian clock in carcinogenesis and anticancer treatments. The understanding of the circadian-cell cycle interaction may contribute to the optimization of drug delivery (see Chap. 6). Robust biomarkers based on chronobiology and systems biology can be used for the establishment of rhythmic profiles toward more precise diagnosis and individualized treatment. Personalized chronotherapy may help improve the treatment of various diseases including hypertension, cancer, depression, and rheumatoid arthritis (see Chap. 7).

By covering topics from cellular networks to complex diseases, from novel concepts to emerging fields, this book intends to provide a state-of-the-art and integrative view of cellular rhythms and networks with potential clinical applications in personalized and systems medicine. Frameworks on the basis of systems biology and chronobiology are introduced for understanding the complexity in health and diseases.

I would like to thank the editors from Springer for their support in this exciting project.

Santa Clara, CA, USA   Qing Yan

Contents

About the Author

Dr. Qing Yan received her Ph.D. in Biological and Medical Informatics from the University of California San Francisco (UCSF). She has extensive research experience in biomedical sciences and leadership experience in the biopharmaceutical industry. Her research interests include bioinformatics, systems biology, pharmacogenomics, immunology, neuroscience, personalized medicine, and translational medicine. Dr. Yan has published many research papers and edited four books on biomedical topics.

Chapter 1
Introduction: Cellular Rhythms and Networks in Systems and Dynamical Medicine

1.1 The Importance of Temporal Factors in Personalized Medicine

There are several key elements in the practice of personalized medicine. The exploration of the "variations" in the responses to pathogenic and therapeutic factors may lead to more accurate diagnosis and the right interventions for "the right people with the right dosages and intensities at the right time" (Yan 2014). Specifically, the "variations" should include not just the differences among individuals. The accuracy should also come from the recognition of temporal and periodic differences in each individual.

Temporal elements such as rhythms and oscillations are the essential properties of biological organisms. With the environmental changes caused by the movements of the sun, the moon, and the earth, endogenous biological clocks provide the mechanisms for adaptation to the constant cycles. These internal clocks enable the organism to make psychophysiological adjustments in accordance with the outside variations, such as having more efficient usages of energy resources (Zhang and Kay 2010). Such adaptive strategies are necessary for health and well-being.

Chronological factors have fundamental roles in physical and chemical processes. Composed of complex interactive networks, various organs in the human body are under the influence of temporal factors, especially the central nervous system (CNS) (Kopec and Carew 2013; Gulsuner et al. 2013). If measured at various temporal scales, the physiological components and functional conditions at one time-point are usually very different from those at another time-point.

For example, the circadian clock is an intrinsic oscillator in charge of the daily rhythms in physiological responses and psychological behaviors (Baggs and Hogenesch 2010). Another obvious example is aging, an evolutionary progression over time with changes at various spatial levels, from cell cycles to cell motility, from genetic expression patterns to physiological complexity (Manor and Lipsitz 2013; Jonker et al. 2013).

© Springer International Publishing Switzerland 2015
Q. Yan, *Cellular Rhythms and Networks: Implications for Systems Medicine*,
SpringerBriefs in Cell Biology, DOI 10.1007/978-3-319-22819-8_1

However, the temporal factors have often been ignored in clinical practice and basic scientific studies. As an example, up to now epidemiologic studies about disease risks have been mostly about the "average" patterns in the spatial context (Zhang et al. 2011). Only recently has such trend been changed to embrace the spatiotemporal factors in the analyses of disease risks.

1.2 Temporal Patterns and Oscillations: A Systems Biology Perspective

Rhythmic physiological and psychological activities are maintained on the basis of complex interactions among components at various spatiotemporal levels. A systems biology perspective is necessary to understand the dynamical patterns and to characterize their functions, targets, and interactions. Figure 1.1 shows a framework for studies in systems and dynamical medicine at various levels in the temporal (e.g., days and seasons) and spatial (e.g., cells and organs) dimensions. Dynamical networks and feedback loops can be analyzed in different hierarchies. To understand the impacts of the temporal factors on health and diseases, nonlinear time-series investigations may provide insights into the system dynamics using different time scales at different spatial levels (Yan 2014).

Analyses at various time scales are important, including nanoseconds, seconds, hours, 24-h cycle, days, weeks, seasons, years, and even decades (see Fig. 1.1). To give an example, the ion channel gating activities may be assessed at the microsecond

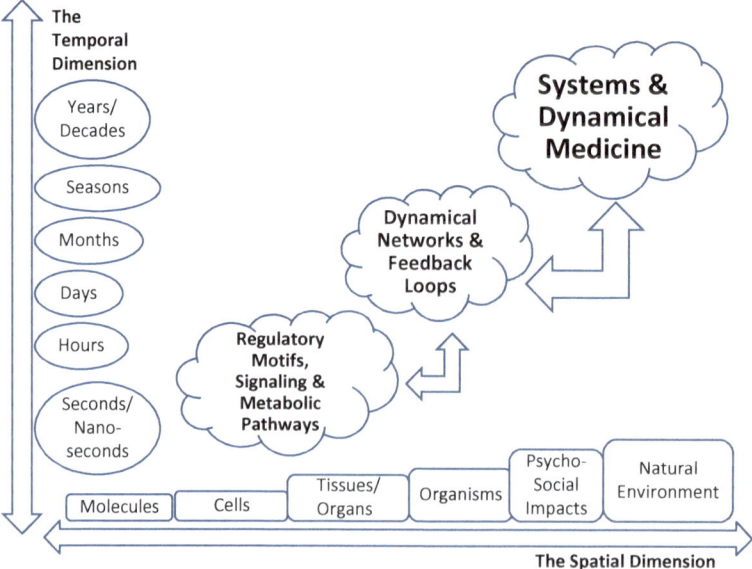

Fig. 1.1 A framework for studies in systems and dynamical medicine at various spatiotemporal levels

scale (Huang and Wikswo 2006). At a higher level, the depolarization of the heart organ may be examined at the millisecond scale. The stability of the cardiac cycle may be tested at the second scale. Moreover, the longevity and aging of the whole organism may be evaluated at the gigasecond scale (Yan 2014).

As shown from these examples, physiological and pathological patterns can be explored with different rhythms and cycles. For instance, circadian clocks running about 24-h play crucial roles in many physiological variations. Up to now among all of the rhythms, the circadian rhythm is the most thoroughly studied type (da Silva Lopes et al. 2013). This book will also focus on the roles of the circadian variations in psychophysiological responses, pathological changes, and therapeutic strategies (see Chaps. 2–7).

Other rhythms are also crucial and deserve more investigations. Specifically, infradian rhythms last longer than 24 h. One relevant physiological example is the menstrual cycle (Halberg et al. 2009; da Silva Lopes et al. 2013). Seasonal cycles often have epidemiological meanings, such as influenza infections in winter. Ultradian rhythms are the cycles that are shorter than 24 h. Many physiological patterns fall into this category, including heartbeats and the firing frequency of neurons. Besides biomedical and ecological rhythms, social cycles also have important impacts. Examples include the weekly working cycle and the cycle of school years. To have a relatively holistic view of the psychosocial and physiological patterns, all of these cycles including biological, environmental, and social factors need to be studied.

Moreover, a systemic understanding of the biomedical rhythms requires the identification of the patterns across different spatial scales from genes to cells to organs. At the molecular and cellular levels, regulatory motifs and metabolic pathways can be identified as potential biomarkers (see Fig. 1.1). For example, genome-wide analyses may help identify the biomolecules and regulatory networks associated with circadian rhythms (De Haro and Panda 2006). In addition, studies about the oscillations in the transcript and protein contexts have pictured cell cycle as an evolving procedure (Klevecz et al. 2008). Periodic dynamics have been identified as the features of the gene expression patterns associated with the cycles of cell division and cellular redox variations (Yan 2014).

Considering cellular organelles, the properties of spatiotemporal dynamics and oscillations have been found in mitochondria and transmembrane potentials (Kurz et al. 2010; Stephane et al. 2012). Looking up to the levels of organs and the whole organism, rhythmic patterns have the essential roles in neural events, brain activities, heart excitation waves, cognitive processes, as well as the working memory (Kurz et al. 2010; Stephane et al. 2012).

Systems biology approaches including gene expression profiling and perturbation assessment may be useful for evaluating the spatiotemporal properties (Zhang and Kay 2010; Baggs and Hogenesch 2010). Proteomic studies, cell-based screening assays, and computational modeling would enable the examination of the feedback loops and clock regulators (Baggs and Hogenesch 2010; Goldbeter et al. 2012). Different oscillatory processes in the cellular rhythms can be explored at different time scales, including the daily circadian pattern, segmentation cycles, and synthetic rhythms (Goldbeter et al. 2012). The oscillations in various networks can also be characterized, such as those of the p53 and NFκB pathways (see Chap. 2). Such discoveries using systems methods may lead to the development of novel diagnostic and therapeutic strategies.

Although this book focuses mainly on the cellular level, an understanding on the basis of systems biology requires the holistic view of the collective or "emergent" features across various spatial levels and temporal scales (Yan 2014). Such cross-scale analyses in all ranges would enable a better understanding of the interrelationships among the oscillating behaviors, health, and diseases (see Fig. 1.1). For instance, during the formation processes of memory, the growth factor (GF) signaling has a critical role in both structural and behavioral plasticity (Kopec and Carew 2013). A more complete understanding of the processes requires the establishment of both spatial and temporal profiles of the GF signaling to describe the interactive elements in the complex network. The integrative profiling approaches covering multiple spatiotemporal scales would allow for the construction of systemic networks to be applied for preventive and precision medicine (Yan 2014).

1.3 Cellular Rhythms and Dynamical Diseases

Studies of biological rhythms at various levels such as the cellular level may have profound implications for health care. The so called "dynamical diseases" have the features of varied dynamical complexity and rhythms that are different from those in the normal conditions (Yan 2014). Many common diseases have already been confirmed to have such properties, including depression, schizophrenia, Parkinson's disease, as well as aging and age-associated disorders (Pezard et al. 1996; An der Heiden 2006; Schiff 2010). In the problem of obesity, a nonlinear association has been found between diurnal cortisol production and the pathogenesis of the disorder (Kumari et al. 2010).

Although depression is categorized in the mental disorder area, it has physiological variables that are featured with oscillations. These variables have nonlinear interrelationships, forming the complexity in the disease (Tretter et al. 2011). Because the nervous system is characterized with "a hierarchy of oscillatory processes" (Milton and Black 1995), such oscillations have profound impacts on various tissues, organs, and systems. On the other hand, altered oscillatory activities of gene expressions have been related to the distorted rhythmic manners in psychiatric disorders (Gebicke-Haerter et al. 2013). More discussions of the connections between circadian rhythms and mental disorders can be found in Chap. 4.

The dynamical property is remarkable in chronic diseases such as cancer (see Chap. 6). For instance, the close relationship between B cell cycles and chronic lymphocytic leukemia (CLL) has made the disorder a typical dynamical disease (Damle et al. 2010). Recent advancements in the research of microRNAs (miRNAs) have identified their close relationships with the regulation of cell cycles. Such correlations emphasize their roles in the complex disorders including cancer and highlight the dynamical features in health and diseases (Stahlhut Espinosa and Slack 2006).

These examples support the fact that "everything oscillates" (Klevecz et al. 2008). Such oscillations can be observed at various levels, from mitochondria to cell cycles, from heartbeats to cognition. Based on such perspectives, all diseases are "dynamical diseases" that require dynamical analyses at every step of medical practice, from diagnostic approaches to therapeutic strategies. Methods in "dynamical medicine" need to be established, such as the investigations of bifurcations and nonlinear time-series analyses (Yan 2014).

Such approaches would enable the establishment of systemic models for psychophysiological and pathological oscillations and feedback loops in order to achieve personalized medicine. For instance, in psychiatrics, models of dynamical systems have been applied to characterize systematic oscillations in the patterns of mental symptoms (Odgers et al. 2009). Biomarkers and symptoms related to temporal complexity and alterations have been studied in a wide spectrum of diseases such as asthma (Frey et al. 2011). In prostate cancer, therapeutic sensitivity and resistance during various stages in its progression have been associated with molecular and cellular elements in the tumor cells (Shaffer and Scher 2003).

Because the dynamical properties have essential roles in health and diseases, it is necessary to recognize the systemic biomarkers as the shifting targets, such as the interactive molecules and cellular pathways at various time points or stages (also see Chap. 7). Such identification would allow for more accurate diagnosis and more effective therapies. In addition, to address the oscillatory features of dynamical diseases, frequent follow-ups are needed to prevent disease progression and potential recurrence. For example in depressive disorders, preventive intervention strategies can be developed for those patients with potential recurrence (Pezard et al. 1996). In the following section, the importance of systems and dynamical biomarkers will be reviewed to elucidate their roles in dynamical diseases.

1.4 Systems and Dynamical Biomarkers

Representing physiological and pathological conditions, biomarkers have significant roles in the diagnosis and prognosis of diseases. As indicators of abnormal functions and treatment responses, biomarkers are useful for the prediction and prevention of the development of diseases. The disease profiling and classification on the basis of biomarkers may also contribute to the discovery of novel therapeutics.

However, because diseases have the complexity and dynamical properties, it is often challenging to accurately characterize and validate biomarkers. Conventional methods for biomarker identification including the symptom checklists often fail to recognize the multifaceted conditions and dynamical stages for precise and timely diagnosis and prognosis (Yan 2014). Even the genetic tests developed in recent decades are mostly based on the detection of single biomolecules or single nucleotide polymorphisms (SNPs) that cannot represent the composite malfunctions or complex networks. In addition to the theoretical aspect, the advancement in technologies also calls for novel approaches in the identification of biomarkers.

For instance, limitations have been discussed in using the pathology-type immuno-histochemical markers (IHCM) in clinical pathology (Abu-Asab et al. 2011).

With the recognition of the importance of the oscillations and networks in complex diseases, methods emphasizing systems and dynamical investigation are needed to identify both precise and robust biomarkers. Systems biology and "omics"-based strategies can be developed for the classification of patient subgroups for individualized medicine. Different factors need to be considered at various levels.

At the molecular level, factors such as genomic variations and epigenetic altera-tions need to be included. At the cellular level, high content phenotypic tests can be used to detect the cellular alterations quantitatively. The profiling of the changes in cells' whole transcriptome or proteome can be developed (Dunn et al. 2010). Libraries of peptides and poly-nucleotides such as siRNA can be established to assess the perturbing elements associated with the proteomic and cellular markers. Relevant cellular pathways and networks may also be analyzed (see Chap. 2). At higher levels, changes in the structure–function correlations, etiologic heterogeneity, psychosocial status, and environmental influences can be incorporated (Filiou and Turck 2011).

In addition to the examinations at various levels in the spatial dimension, bio-markers with predictive powers need to be discovered to represent the robust status and disease stages in the temporal dimension (see Fig. 1.1). For example, biomark-ers describing the temporal evolutionary stages have been applied to evaluate the clinical initiation and progression of symptoms in Alzheimer's disease (AD) (Jack et al. 2013). Dynamical models may address the temporal factors including cellular rhythmic networks for depicting disease progressions, which are more accurate than using the clinical symptom severity as biomarkers.

The advancement of technologies in recent years may enable the cross-scale investigations at various spatiotemporal levels. Specifically, high-throughput (HTP) technologies can be applied for the dynamical assessments of systems-wide geno-type–phenotype correlations with time-series examinations (Chen et al. 2012). Before using such technologies, the traditional genetic markers could only differen-tiate disease from healthy samples in static conditions that may lead to imprecise drug design and adverse clinical outcomes. However, the new technologies may help avoid these inadequacies.

As an example, the relapses of complex diseases often happen suddenly at a tip-ping point with a forthcoming bifurcation. The capture of this critical stage with the recognition of early-warning signals would be meaningful for early diagnosis and prognosis toward preventive medicine. The HTP gene expression methods have been suggested useful for the detection of such dynamical network biomarkers (DNBs) (Chen et al. 2012). Tissue-specific molecular and cellular markers in DNBs can be determined for the transition from the normal to disease conditions (Li et al. 2014).

For instance, cancer has progressive stages featured with dynamical patterns of proteomic changes and shift tendencies of carcinogenesis. Such properties make early detection and interventions as the optimal method to lower the mortality of solid cancers (Li et al. 2011). Conventionally, cancer screening models have classified at-risk people into three groups, i.e., normal people, those without symptoms, and patients with detectable symptoms. However, such static categorization has not been helpful for reducing the mortality rate. To make improvements, models such as those embracing the dynamic clonal evolution can be very useful for systemic biomarker discovery (Li et al. 2011).

In summary, the discovery of the clusters of robust biomarkers such as cellular rhythmic networks may help improve the accuracy in the risk identification and the prediction of disease progression. Such approaches may allow for the stratification of different at-risk patient groups for timely therapies and individualized care that are essential in personalized medicine. More discussions on systems and dynamical biomarkers for various diseases will be available in Chap. 7.

1.5 Bioinformatics Support for Spatiotemporal Studies

The development of systems biology relies on both experimental and computational approaches. The cross-scale dynamical analyses for spatiotemporal studies may become the basis for the practice of personalized, predictive, and preventive medicine (Yan 2014). An indispensable method toward the achievement of these goals is bioinformatics. For instance, mathematical and computational models have been found useful to simulate neuron–neuron communications, neuron–glia gap junctions, and various neurobehavioral functions (Kronauer et al. 2007; Gebicke-Haerter et al. 2013). Laboratory tests and computational algorithms can be designed for time-series analyses of the cellular rhythmic networks and individual differences in such complex processes.

Figure 1.2 provides a schematic summary of the bioinformatics approaches to support the development of systems and dynamical medicine. By using methods of data integration, data mining, knowledge discovery, and decision support, systems models can be established for the identification of dynamical biomarkers based on the spatiotemporal profiles. Such approaches would enable the understanding of the mechanisms at various levels, including the analyses in functional genomics, proteomics, epigenomics, pharmacogenomics, chronobiology, and systems biology. On the basis of the systems models and spatiotemporal profiles, more accurate diagnosis and prognosis may be achieved for predictive, preventive, and personalized medicine.

Many bioinformatics resources are available for the assessment of genomic and proteomic dynamics, as shown in the examples in Table 1.1. For instance, DNAtraffic is an annotated database about protein structures, functions, and DNA damage response pathways (Kuchta et al. 2012). The database can be used to study genome dynamics and DNA networks during the cell life. The Membrane Builder in the CHARMM-GUI website can be used for molecular dynamics and mechanics simulations (Jo et al. 2007). CellFinder provides an ontological platform for genotype–phenotype analyses in various cell types (Seltmann et al. 2013). It can be used to characterize cells in different developmental stages and their roles in tissues, organs and organisms. Cyclebase is a database about cell-cycle regulation (Santos et al. 2015).

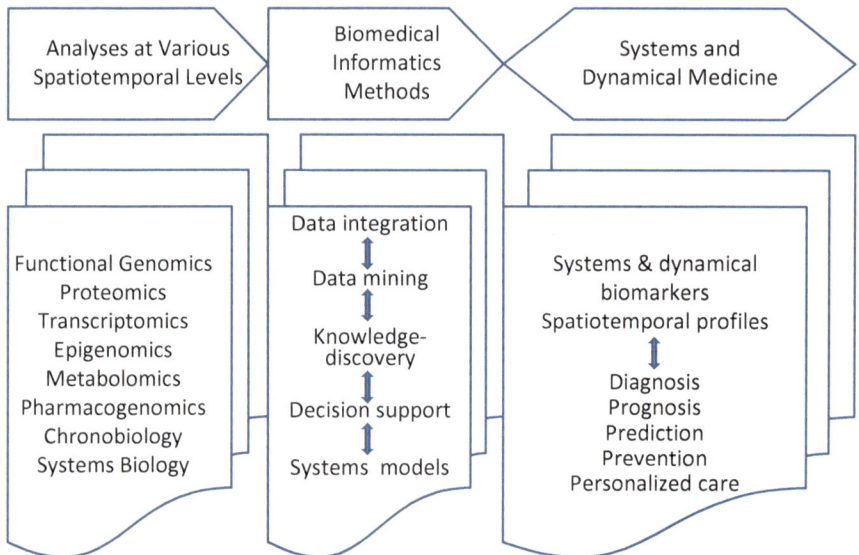

Fig. 1.2 Bioinformatics support for systems and dynamical medicine

Another site, Arena3D hosts a visualization tool for the analyses of the time-driven phenotypic patterns and morphological layers (Secrier et al. 2012).

The Conformational Dynamics Data Bank (CDDB) is about the dynamics of proteins with emphasis on the conformational transitions (Kim et al. 2011; also see Table 1.1). Dynameomics is for the dynamics simulation of protein folding and pathways (Van der Kamp et al. 2010). The Dynamic Proteomics is about the dynamics of endogenous proteins in living human cells including protein dynamics fluorescence movies (Frenkel-Morgenstern et al. 2010). The Eurexpress atlas is a database and map of the transcriptome in the mouse embryo for the analysis of spatiotemporal gene expression patterns (Diez-Roux et al. 2011). The MitoGenesisDB is a database about the spatiotemporal dynamics of mitochondrial protein formation including the time-course of mRNA generation (Gelly et al. 2011).

At the tissue and organ levels, the Allen Brain Atlas is for spatiotemporal studies of the CNS including genome-wide maps (Sunkin et al. 2013; also see Table 1.1). The Cerebellar Development Transcriptome Database (CDT-DB) is about spatiotemporal gene expression patterns in mouse cerebellar development (Sato et al. 2008). EpiScanGIS utilizes a Geographical Information System (GIS) for the mapping of the spatiotemporal data of meningococcal disease (Reinhardt et al. 2008).

In addition, The EUCLOCK Information System (EUCLIS) is about circadian biology and chronobiology (Batista et al. 2007). DBBR is a database to support systems biology studies in biological rhythms (DBBR 2015; also see Table 1.1). CircaDB provides a collection of circadian gene expression data based on experiments (Pizarro et al. 2013). CircadiOmics is about circadian genomics, transcriptomics, and metabolomics (Patel et al. 2012).

Table 1.1 Examples of bioinformatics resources for spatiotemporal studies

Tools	URLs[a]	Contents
Allen Brain Atlas	http://www.brain-map.org	A spatiotemporal platform for the CNS
Arena3D	http://arena3d.org	Time-driven phenotypes
CellFinder	http://cellfinder.org	Systematic cell types
CircaDB	http://bioinf.itmat.upenn.edu/circa/	Circadian gene expressions
CircadiOmics	http://circadiomics.igb.uci.edu/	Circadian genomics, transcriptomics, metabolomics
CHARMM-GUI	http://www.charmm-gui.org/	Macromolecular dynamics
CDDB	http://www.cdyn.org/	Protein conformational dynamics
CDT-DB	http://www.cdtdb.neuroinf.jp/CDT/Top.jsp	Spatiotemporal mouse brain gene expressions
Cyclebase	http://www.cyclebase.org/CyclebaseSearch	Cell-cycle regulation
DBBR	http://pharmtao.com/health/biological-rhythms-database/	Systems biology of biological rhythms
DNAtraffic	http://dnatraffic.ibb.waw.pl/	Genome dynamics
Dynameomics	http://www.dynameomics.org	Protein dynamics
Dynamic Proteomics	http://www.weizmann.ac.il/mcb/UriAlon/DynamProt/	Protein dynamics
EpiScanGIS	http://www.episcangis.org	Spatiotemporal disease clusters
EUCLIS	http://www.bioinfo.mpg.de/euclis/	Circadian biology
Eurexpress atlas	http://www.eurexpress.org	Mouse embryo transcriptomes
MitoGenesisDB	http://www.dsimb.inserm.fr/dsimb_tools/mitgene/	Spatiotemporal mitochondrial dynamics

[a]Resources accessed May 15, 2015

1.6 Conclusion: Systems and Dynamical Medicine

The objectives of personalized medicine require systemic and cross-scale analyses of spatiotemporal patterns, from molecules to cells and organisms, from seconds to days and years. On the basis of such understanding, the development of systems and dynamical medicine addressing timely changes in the whole systems would be possible.

While the term "systems medicine" emphasizes the perception of holism, the phrase of "dynamical medicine" underscores the adaptations of the interwoven spatio-temporal elements in the complex biological systems (see Fig. 1.1). Specifically, the analyses of the nonlinearity and interconnectivity in cellular rhythmic networks and feedback loops may contribute to a more integrative and proactive care in the clinic.

For example, at the cellular level, the nonlinear dynamical behavior of mito-chondria plays a critical role in controlling energy metabolism within liver cells (Ramanujan and Herman 2007). Variations in this process have been associated with the mechanisms of aging and abnormal functions of various organs.

Going up to the organ level, the examinations of the nonlinear dynamics of heart rates have revealed the importance of multifaceted factors such as the circadian clocks and age (Vandeput et al. 2012).

In conclusion, the investigation of biological rhythms would bring a better insight into dynamical diseases and ultimately dynamical medicine. Many factors are involved in dynamical diseases, from genetic mutations to epigenetic and environmental changes. The incorporation of these factors from a wide spectrum is necessary to allow for a more comprehensive understanding of health and diseases. Such spatiotemporal understanding can then be translated into the clinical practice of personalized and preventive medicine.

References

Abu-Asab MS, Chaouchi M, Alesci S, Galli S, Laassri M, Cheema AK, Atouf F, VanMeter J, Amri H (2011) Biomarkers in the age of omics: time for a systems biology approach. OMICS 15:105–112

An der Heiden U (2006) Schizophrenia as a dynamical disease. Pharmacopsychiatry 39(Suppl 1): S36–S42

Baggs JE, Hogenesch JB (2010) Genomics and systems approaches in the mammalian circadian clock. Curr Opin Genet Dev 20:581–587

Batista RTB, Ramirez DB, Santos RD, del Rosario MCI, Mendoza ER (2007) EUCLIS—an information system for circadian systems biology. IET Syst Biol 1:266–273

Chen L, Liu R, Liu Z-P, Li M, Aihara K (2012) Detecting early-warning signals for sudden deterioration of complex diseases by dynamical network biomarkers. Sci Rep 2:342

da Silva Lopes R, Resende NM, Honorio-França AC, França EL (2013) Application of bioinformatics in chronobiology research. Sci World J 2013:153839

Damle RN, Calissano C, Chiorazzi N (2010) Chronic lymphocytic leukaemia: a disease of activated monoclonal B cells. Best Pract Res Clin Haematol 23:33–45

DBBR (2015) The database of biological rhythms. http://pharmtao.com/health/biological-rhythms-database/. Accessed 1 June 2015

De Haro L, Panda S (2006) Systems biology of circadian rhythms: an outlook. J Biol Rhythms 21:507–518

Diez-Roux G, Banfi S, Sultan M, Geffers L, Anand S, Rozado D, Magen A, Canidio E, Pagani M, Peluso I et al (2011) A high-resolution anatomical atlas of the transcriptome in the mouse embryo. PLoS Biol 9:e1000582

Dunn DA, Apanovitch D, Follettie M, He T, Ryan T (2010) Taking a systems approach to the identification of novel therapeutic targets and biomarkers. Curr Pharm Biotechnol 11:721–734

Filiou MD, Turck CW (2011) General overview: biomarkers in neuroscience research. Int Rev Neurobiol 101:1–17

Frenkel-Morgenstern M, Cohen AA, Geva-Zatorsky N, Eden E, Prilusky J, Issaeva I, Sigal A, Cohen-Saidon C, Liron Y, Cohen L et al (2010) Dynamic proteomics: a database for dynamics and localizations of endogenous fluorescently-tagged proteins in living human cells. Nucleic Acids Res 38:D508–D512

Frey U, Maksym G, Suki B (2011) Temporal complexity in clinical manifestations of lung disease. J Appl Physiol 110:1723–1731

Gebicke-Haerter PJ, Pildaín LV, Matthäus F, Schmitt A, Falkai P (2013) Circadian rhythms investigated on the cellular and molecular levels. Pharmacopsychiatry 46(Suppl 1):S22–S29

Gelly J-C, Orgeur M, Jacq C, Lelandais G (2011) MitoGenesisDB: an expression data mining tool to explore spatio-temporal dynamics of mitochondrial biogenesis. Nucleic Acids Res 39:D1079–D1084

Goldbeter A, Gérard C, Gonze D, Leloup J-C, Dupont G (2012) Systems biology of cellular rhythms. FEBS Lett 586:2955–2965

Gulsuner S, Walsh T, Watts AC, Lee MK, Thornton AM, Casadei S, Rippey C, Shahin H, Consortium on the Genetics of Schizophrenia (COGS), PAARTNERS Study Group et al (2013) Spatial and temporal mapping of de novo mutations in schizophrenia to a fetal prefrontal cortical network. Cell 154:518–529

Halberg F, Cornélissen G, Wilson D, Singh RB, De Meester F, Watanabe Y, Otsuka K, Khalilov E (2009) Chronobiology and chronomics: detecting and applying the cycles of nature. Biologist (London) 56:209–214

Huang S, Wikswo J (2006) Dimensions of systems biology. Rev Physiol Biochem Pharmacol 157:81–104

Jack CR, Knopman DS, Jagust WJ, Petersen RC, Weiner MW, Aisen PS, Shaw LM, Vemuri P, Wiste HJ, Weigand SD et al (2013) Tracking pathophysiological processes in Alzheimer's disease: an updated hypothetical model of dynamic biomarkers. Lancet Neurol 12:207–216

Jo S, Kim T, Im W (2007) Automated builder and database of protein/membrane complexes for molecular dynamics simulations. PLoS One 2:e880

Jonker MJ, Melis JPM, Kuiper RV, van der Hoeven TV, Wackers PFK, Robinson J, van der Horst GTJ, Dollé MET, Vijg J, Breit TM et al (2013) Life spanning murine gene expression profiles in relation to chronological and pathological aging in multiple organs. Aging Cell 12:901–909

Kim D-N, Altschuler J, Strong C, McGill G, Bathe M (2011) Conformational dynamics data bank: a database for conformational dynamics of proteins and supramolecular protein assemblies. Nucleic Acids Res 39:D451–D455

Klevecz RR, Li CM, Marcus I, Frankel PH (2008) Collective behavior in gene regulation: the cell is an oscillator, the cell cycle a developmental process. FEBS J 275:2372–2384

Kopec AM, Carew TJ (2013) Growth factor signaling and memory formation: temporal and spatial integration of a molecular network. Learn Mem 20:531–539

Kronauer RE, Gunzelmann G, Van Dongen HPA, Doyle FJ, Klerman EB (2007) Uncovering physiologic mechanisms of circadian rhythms and sleep/wake regulation through mathematical modeling. J Biol Rhythms 22:233–245

Kuchta K, Barszcz D, Grzesiuk E, Pomorski P, Krwawicz J (2012) DNAtraffic—a new database for systems biology of DNA dynamics during the cell life. Nucleic Acids Res 40:D1235–D1240

Kumari M, Chandola T, Brunner E, Kivimaki M (2010) A nonlinear relationship of generalized and central obesity with diurnal cortisol secretion in the Whitehall II study. J Clin Endocrinol Metab 95:4415–4423

Kurz FT, Aon MA, O'Rourke B, Armoundas AA (2010) Spatio-temporal oscillations of individual mitochondria in cardiac myocytes reveal modulation of synchronized mitochondrial clusters. Proc Natl Acad Sci U S A 107:14315–14320

Li X, Blount PL, Vaughan TL, Reid BJ (2011) Application of biomarkers in cancer risk management: evaluation from stochastic clonal evolutionary and dynamic system optimization points of view. PLoS Comput Biol 7:e1001087

Li M, Zeng T, Liu R, Chen L (2014) Detecting tissue-specific early warning signals for complex diseases based on dynamical network biomarkers: study of type 2 diabetes by cross-tissue analysis. Brief Bioinform 15:229–243

Manor B, Lipsitz LA (2013) Physiologic complexity and aging: implications for physical function and rehabilitation. Prog Neuropsychopharmacol Biol Psychiatry 45:287–293

Milton J, Black D (1995) Dynamic diseases in neurology and psychiatry. Chaos 5:8–13

Odgers CL, Mulvey EP, Skeem JL, Gardner W, Lidz CW, Schubert C (2009) Capturing the ebb and flow of psychiatric symptoms with dynamical systems models. Am J Psychiatry 166:575–582

Patel VR, Eckel-Mahan K, Sassone-Corsi P, Baldi P (2012) CircadiOmics: integrating circadian genomics, transcriptomics, proteomics and metabolomics. Nat Methods 9:772–773

Pezard L, Nandrino JL, Renault B, el Massioui F, Allilaire JF, Müller J, Varela F, Martinerie J (1996) Depression as a dynamical disease. Biol Psychiatry 39:991–999

Pizarro A, Hayer K, Lahens NF, Hogenesch JB (2013) CircaDB: a database of mammalian circadian gene expression profiles. Nucleic Acids Res 41:D1009–D1013

Ramanujan VK, Herman BA (2007) Aging process modulates nonlinear dynamics in liver cell metabolism. J Biol Chem 282:19217–19226

Reinhardt M, Elias J, Albert J, Frosch M, Harmsen D, Vogel U (2008) EpiScanGIS: an online geographic surveillance system for meningococcal disease. Int J Health Geogr 7:33

Santos A, Wernersson R, Jensen LJ (2015) Cyclebase 3.0: a multi-organism database on cell-cycle regulation and phenotypes. Nucleic Acids Res 43:D1140–D1144

Sato A, Sekine Y, Saruta C, Nishibe H, Morita N, Sato Y, Sadakata T, Shinoda Y, Kojima T, Furuichi T (2008) Cerebellar development transcriptome database (CDT-DB): profiling of spatio-temporal gene expression during the postnatal development of mouse cerebellum. Neural Netw 21:1056–1069

Schiff SJ (2010) Towards model-based control of Parkinson's disease. Philos Trans A Math Phys Eng Sci 368:2269–2308

Secrier M, Pavlopoulos GA, Aerts J, Schneider R (2012) Arena3D: visualizing time-driven phenotypic differences in biological systems. BMC Bioinf 13:45

Seltmann S, Stachelscheid H, Damaschun A, Jansen L, Lekschas F, Fontaine J-F, Nguyen-Dobinsky TN, Leser U, Kurtz A (2013) CELDA – an ontology for the comprehensive representation of cells in complex systems. BMC Bioinf 14:228

Shaffer DR, Scher HI (2003) Prostate cancer: a dynamic illness with shifting targets. Lancet Oncol 4:407–414

Stahlhut Espinosa CE, Slack FJ (2006) The role of microRNAs in cancer. Yale J Biol Med 79:131–140

Stephane M, Leuthold A, Kuskowski M, McClannahan K, Xu T (2012) The temporal, spatial, and frequency dimensions of neural oscillations associated with verbal working memory. Clin EEG Neurosci 43:145–153

Sunkin SM, Ng L, Lau C, Dolbeare T, Gilbert TL, Thompson CL, Hawrylycz M, Dang C (2013) Allen Brain Atlas: an integrated spatio-temporal portal for exploring the central nervous system. Nucleic Acids Res 41:D996–D1008

Tretter F, Gebicke-Haerter PJ, an der Heiden U, Rujescu D, Mewes HW, Turck CW (2011) Affective disorders as complex dynamic diseases—a perspective from systems biology. Pharmacopsychiatry 44(Suppl 1):S2–S8

Van der Kamp MW, Schaeffer RD, Jonsson AL, Scouras AD, Simms AM, Toofanny RD, Benson NC, Anderson PC, Merkley ED, Rysavy S et al (2010) Dynameomics: a comprehensive database of protein dynamics. Structure 18:423–435

Vandeput S, Verheyden B, Aubert AE, Van Huffel S (2012) Nonlinear heart rate dynamics: circadian profile and influence of age and gender. Med Eng Phys 34:108–117

Yan Q (2014) From pharmacogenomics and systems biology to personalized care: a framework of systems and dynamical medicine. Methods Mol Biol 1175:3–17

Zhang EE, Kay SA (2010) Clocks not winding down: unravelling circadian networks. Nat Rev Mol Cell Biol 11:764–776

Zhang Z, Chen D, Liu W, Racine JS, Ong S, Chen Y, Zhao G, Jiang Q (2011) Nonparametric evaluation of dynamic disease risk: a spatio-temporal kernel approach. PLoS One 6:e17381

Chapter 2
Circadian Rhythms and Cellular Networks: A Systems Biology Perspective

2.1 Systems Biology of Circadian Rhythms: A Schematic Overview

The emerging field of systems biology represents the switch of the gear from reductionist approaches toward the recognition and understanding of complex systems and behaviors. Systems biology emphasizes the interactions among individual elements such as genes and proteins. Such focus makes the circadian timekeeping systems a valuable model. This is because the circadian oscillating systems have the features of robustness, periodicity, nonlinearity, adaptation, temperature compensation, as well as synchronization (Ueda 2007; Hogenesch and Ueda 2011). As an essential adaptive function, the organism's internal physiological and behavioral rhythms can be entrained and synchronized to environmental signals.

Figure 2.1 provides a schematic overview of the roles of the clock systems in various psychophysiological processes at different levels. Specifically, the environment (e.g., the light/dark cycle) and lifestyles (e.g., sleep patterns) may influence the central and peripheral clocks. While the suprachiasmatic nucleus (SCN) of the hypothalamus harbors the master clock, the peripheral clocks in the heart, lung, liver, kidney, and muscles also regulate the behavioral and physiological activities. The central SCN clock may orchestrate the circadian oscillators in peripheral tissues with light signals as the foremost synchronizer.

In addition, various rhythmic signals from the hypothalamic–pituitary–adrenal (HPA) axis and immune–endocrine systems such as hormones (e.g., melatonin) and cytokines (e.g., TGFβ) also participate in the oscillations. At the molecular and cellular levels, the circadian genes and pathways interact with multiple cellular networks in the regulation of cell cycle, transcription, metabolism, electrical activities, DNA repair, and stress responses (see Fig. 2.1). More details of these processes and their relevant pathogenesis will be discussed in Chaps. 3–7.

As illustrated in Fig. 2.1, the 24-h cycle systems are critical for understanding the biomedical complexity across various spatiotemporal scales (see Chap. 1).

© Springer International Publishing Switzerland 2015
Q. Yan, *Cellular Rhythms and Networks: Implications for Systems Medicine*,
SpringerBriefs in Cell Biology, DOI 10.1007/978-3-319-22819-8_2

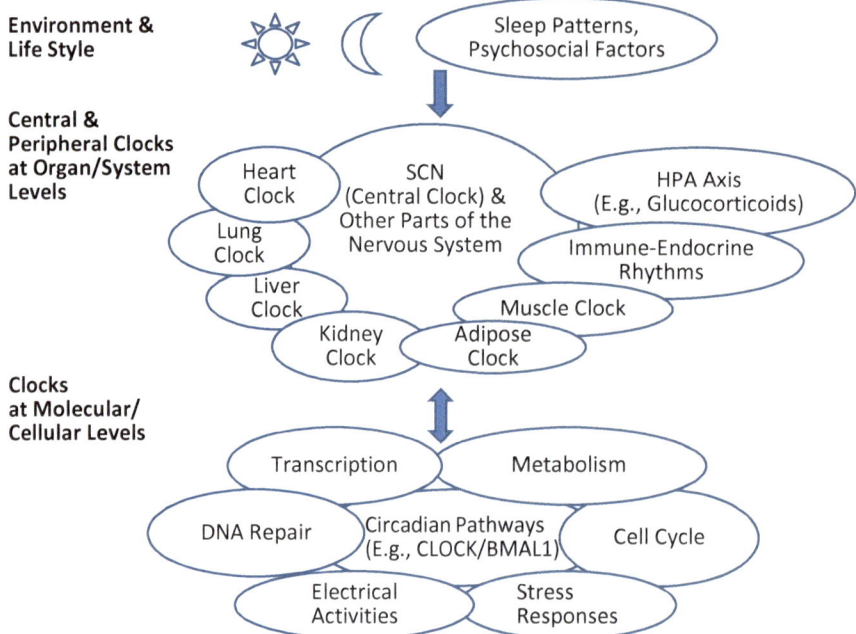

Fig. 2.1 The roles of the clock systems in psychophysiological activities at different levels

The endogenous oscillators control daily rhythms from cis elements to genes, from transcriptional circuits to cellular pathways, from the SCN network to behaviors in various timeframes (Yamada and Forger 2010; Ukai and Ueda 2010; Baggs and Hogenesch 2010). Such intrinsic clocks allow for the adaptation to environmental changes and efficient utilization of energy sources (Zhang and Kay 2010). The adaptive advantages are pivotal for health maintenance and recovery from illnesses.

At the molecular and cellular levels, biological rhythms come from the feedback loops generated in the regulatory networks. Table 2.1 shows some examples of the proteins with core roles in the regulation of the circadian clocks. The relevant interactions and diseases are also included. A more complete list can be found in the Database of Biological Rhythms (DBBR 2015). For example, the protein PER1 is an essential member of the PER/CRY and CLOCK/BMAL1 feedback loops. It is critical for circadian-associated locomotor activity, metabolism, and behaviors. Abnormal functions of PER1 have been associated with pancreatic cancer (Sato et al. 2009), prostate cancer (Cao et al. 2009), buccal squamous cell carcinoma (BSCC) (Zhao et al. 2013), and skin tumorigenesis (Lengyel et al. 2013) (see Table 2.1).

Studies of such rhythmic proteins and networks would help elucidate a spectrum of oscillatory processes including those of p53 and NFκB, as well as the associations between the dynamics of cyclin-dependent kinases and cell cycle (Goldbeter et al. 2012; also see Sects. 2.2 and 2.3). Furthermore, cell-based screening and proteomics examinations would enable the identification of novel

Table 2.1 Examples of circadian-associated proteins and relevant diseases

Core proteins (interactions)	Roles in circadian rhythms	Associated diseases (references)
ARNTL (also known as BMAL1) (CLOCK/BMAL1, PER/CRY)	Circadian feedback loop	Atherosclerosis (Lin et al. 2014); bipolar disorder (Rybakowski et al. 2014); colorectal cancer (Zeng et al. 2014); gestational diabetes mellitus (Pappa et al. 2013)
CLOCK (CLOCK/BMAL1, PER/CRY)	Circadian regulation	Breast cancer (Truong et al. 2014); obesity (Dashti et al. 2015); sleep quality (Vanderlind et al. 2014)
CRY1 (CLOCK/BMAL1, PER/CRY)	Circadian regulation	Chronic lymphocytic leukemia (Lewintre et al. 2009); colorectal cancer (Yu et al. 2013); glioma (Luo et al. 2012); major depressive disorder (Hua et al. 2014)
CRY2 (CLOCK/BMAL1, PER/CRY)	Circadian regulation	Bipolar type 1 disorder, winter depression (Kovanen et al. 2013); cancer (Hoffman et al. 2010); rapid cycling in bipolar disorder (Sjöholm et al. 2010); type 2 diabetes (Liu et al. 2011)
NR1D1 (BMAL1/CLOCK)	Metabolic, inflammatory and cardiovascular processes	Bipolar disorder (Partonen 2012); mood disorders (Kishi et al. 2008); mycobacterium tuberculosis (Chandra et al. 2013)
PER1 (CLOCK/BMAL1, PER/CRY)	Circadian associated locomotor activity, metabolism, behaviors	Buccal squamous cell carcinoma (BSCC) (Zhao et al. 2013); pancreatic cancer (Sato et al. 2009); prostate cancer (Cao et al. 2009); skin tumor (Lengyel et al. 2013)
PER2 (CLOCK/BMAL1, PER/CRY)	Circadian associated locomotor activity, metabolism, behaviors	Colorectal carcinoma (Štorcelová et al. 2013); familial advanced sleep-phase disorder (FASPD) (Chong et al. 2012)
PER3 (CLOCK/BMAL1, PER/CRY)	Circadian associated locomotor activity, metabolism, behaviors	Non-small cell lung cancer (NSCLC) (Couto et al. 2014); sleep-loss-related attentional lapses (Maire et al. 2014)
TIMELESS (PER1, PER2, PER3, CLOCK/BMAL1)	Cell survival, DNA polymerase epsilon activity, telomere length, epithelial cell morphogenesis	Bipolar mood disorder (Rybakowski et al. 2014); breast tumor (Fu et al. 2012); lung cancer (Yoshida et al. 2013)

clock components and modifiers (Baggs and Hogenesch 2010). The multi-scale view of circadian rhythms on the basis of systems biology would empower the discovery of novel therapeutic strategies such as chronotherapy (see Chap. 7).

2.2 Circadian Rhythms, Protein–Protein Interactions, and Cellular Networks

As the internal timekeepers, the circadian clocks enable the organisms to adapt to the cyclic environmental fluctuations of light and temperature (Schöning and Staiger 2005). They form the essential cellular timing mechanisms that synchronize vital physiological processes. The major molecular clock elements have periodic expression patterns that are driven by cell-autonomous transcriptional feedback loops. Such oscillation is kept in the central clockwork and transmitted to the downstream genes. Post-transcriptional and post-translational processes are critical for maintaining the normal functions of the clock proteins that interact with other proteins. In another word, the circadian rhythm circuitry depends on the interlocked transcription–translation feedback loops with multiple molecules and complex protein–protein interactions involved.

One of such loops is a positive feedback loop operated by the CLOCK/BMAL1 heterodimer that may start the transcription of target genes with elements of the E-box cis-regulatory enhancer sequences (Ko and Takahashi 2006). In addition, a negative feedback loop contains the rhythmic transcription of the clock genes PER1, PER2, and PER3, as well as the cryptochrome genes CRY1 and CRY2. The combination of PER and CRY proteins can construct a heterodimer that may interact with the CLOCK/BMAL1 heterodimer to inhibit its own transcription (Ko and Takahashi 2006). PER and CRY proteins can be phosphorylated by casein kinase epsilon (CKIepsilon) and result in the starting over of the cycle. Another regulatory loop involves the interactions between the CLOCK/BMAL1 heterodimers and the retinoic acid-related orphan nuclear receptors (RORs). While RORs may activate the transcription of BMAL1, REV-ERBs (also known as NR1Ds) may inhibit the transcription process, forming both positive and negative regulations (Ko and Takahashi 2006).

The activation of several signal transduction cascades may be involved in the circadian regulation, such as the cAMP signaling pathways and mitogen-activated protein kinase (MAPK) signaling pathways (Zhang et al. 2010a; Goldsmith and Bell-Pedersen 2013). These pathways are associated with essential physiological functions such as hepatic gluconeogenesis and stress responses (see Table 2.2). Another important phase shifting-associated component is melatonin that may provide the feedback of the immune responses on circadian timing (Fernandes et al. 2006).

Table 2.2 lists some examples of the protein–protein interactions (PPIs) and cellular pathways associated with the regulation of circadian rhythms. The relevant diseases are also included. A more complete list can be found from the Database of Biological Rhythms (DBBR 2015). The processes and timing of the PPIs are essential for all biological activities and normal regulatory functions. Using systems

Table 2.2 Examples of circadian-associated pathways and relevant disorders

Pathways (interactions)	Circadian associations	Relevant disorders	References
AHR signaling pathway (AHR, BMAL1)	Behaviors, cell proliferation, circadian disruptions	Disorders in behaviors, metabolism, immune responses	Shimba and Watabe (2009)
cAMP signaling pathways (CRY1)	Circadian regulation of gluconeogenesis	Type 2 diabetes	Zhang et al. (2010a)
Dopamine signaling (PER2)	Circadian and interval timing	Impaired interval timing and time perception	Bussi et al. (2014)
Leptin and CCK pathways	Circadian regulation of energy balance	Disorders in food intake and energy balance	Merino et al. (2008)
Leptin signaling pathway (SOCS3, JAK)	Circadian patterns, macronutrient storage, energy balance	Metabolic disorders	Ptitsyn and Gimble (2007)
Leptin signaling pathway (STAT-3)	Circadian rhythms of metabolic activity	Obesity, physical inactivity, higher food intake	Hsuchou et al. (2013)
Melatonin synthetic pathway (TNFα)	Circadian timing in immune responses	Inflammatory responses	Fernandes et al. (2006)
Mitogen-activated protein kinase (MAPK) pathways	Clock cycling	Stress responses	Goldsmith and Bell-Pedersen (2013)
mTOR pathway (Fbxw7)	Circadian oscillations of mTOR activity	Cancer, renal cell carcinoma	Okazaki et al. (2014)
mTOR/4E-BP1 pathway (VIP)	SCN entrainment, synchrony	Circadian disruptions	Cao et al. (2013)
NFκB pathway	NFκB oscillation	Inflammatory responses	Wang et al. (2015)
NFκB pathway (CRY, p53, TNFα, GSK3β kinase)	Apoptosis	Cancers	Lee and Sancar (2011)
REV-ERBα (NR1D1) pathway (NFκB, NrF2)	Circadian regulation of cellular metabolism	Lung inflammation, oxidative stress	Yang et al. (2014)
ARNTL (BMAL1)-p53 pathway (PER2)	Circadian dysfunctions	Cancer	Mullenders et al. (2009)
p53 signaling pathway (PER2)	DNA damage responses	Genotoxic stress responses	Gotoh et al. (2015)
SIRT1-BMAL1 pathway	Tobacco/cigarette smoke caused circadian disruption	Chronic obstructive pulmonary disease (COPD), lung inflammation/injury	Hwang et al. (2014)
Wnt/beta-catenin pathway (PPAR gamma-BMAL1)	Cardiovascular rhythms, circadian variations in BP and heart rate	Cardiac dysfunction, arrhythmogenic right ventricular cardiomyopathy, type 2 diabetes	Lecarpentier et al. (2014)
Wnt pathway (β-catenin, BMAL1)	Circadian and cell proliferation	Premature aging	Lin et al. (2013)
Wnt signaling pathway	Circadian disruptions	Cancer, tumorigenesis, tumor growth	Yasuniwa et al. (2010)

biology approaches and systematic circadian phenotyping, recent studies have found that dynamic circadian PPIs and networks are the key connections among various cellular processes from signal transductions to cell cycles (Wallach et al. 2013).

As shown in the examples in Table 2.2, the temporal regulations of cellular physiology and pathology rely on these dynamical networks. For instance, the proteins CRY, p53, TNFα, NFκB, and GSK3β kinase may be involved in the NFκB signaling pathways associated with apoptosis. Abnormal functions in these interactive networks have been related to cancers (Lee and Sancar 2011; also see Chap. 6). In another example, the leptin signaling pathway is involved in the circadian regulation of metabolic activities with malfunctions related to metabolic disorders and obesity (Hsuchou et al. 2013; Merino et al. 2008).

2.3 Two Essential Cellular Rhythms: The Circadian–Cell Cycle Interactions

The two essential cellular rhythms are the cell division cycle and the 24-h circadian pattern. These two coupled oscillators are tightly connected in multiple manners. The 'gating' controls of the circadian systems at different checkpoints of the cell cycle and the impact of cell cycle on the biological rhythms indicate that these intertwined bidirectional circuits are critical in physiological and pathological processes (Masri et al. 2013). Multiple regulatory steps and complex feedback loops are involved to warrant such timekeeping. Studies using temperature, genetic, and pharmacological perturbations have suggested that the two interacting cellular oscillators may adopt robust synchronization over a broad range of parameters (Bieler et al. 2014). The circadian regulation of the cell cycle may cover various stages including S, G1 and G1/S, G2 and G2/M (Weigl et al. 2013). The G1/S transition may also influence the local clocks in the proliferating tissues.

Many molecular elements of the cell cycle network are regulated in a circadian way (Gérard and Goldbeter 2012; El Cheikh et al. 2014). The circadian clock is involved in the regulation of the cell cycle and check-point-associated proteins, which in turn also participate in the regulation of the circadian proteins. For instance, the network of cyclin-dependent kinases (CDKS) controls the development of the consecutive stages of the cell cycle (Gérard and Goldbeter 2012). In this network, the generation of the kinase WEE1 that suppresses the G2/M transition can be promoted by CLOCK-BMAL1, the complex essential in the circadian rhythm network. Another element in the circadian network, REV-ERBα (also known as NR1D1), may suppress the production of the CDK inhibitor p21. The CLOCK-BMAL1 complex may also inhibit the generation of the oncogene c-MYC, while c-MYC may enhance the production of G1 cyclin (Gérard and Goldbeter 2012).

In addition, the multifunctional nuclear protein NONO serves as not only a partner of the circadian PER proteins, but also the linkage between the circadian gating and the cell cycle. Such connection has been found essential for wound healing in mice (Kowalska et al. 2013). NONO may interact with the p16-Ink4A cell cycle checkpoint gene and be involved in the circadian activation that is PER-dependent.

On the other hand, this activation as well as the circadian cell cycle gating may not happen with the depletion of NONO or PER (Kowalska et al. 2013). The loss of NONO may lead to defective wound repair. Such effects indicate that NONO may have a key role in the coupling of the cell cycle to the circadian clock.

The circadian–cell cycle interactions may also affect the cell population growth rates. The coupling of the circadian clock and cell cycle may be mediated via the protein WEE1 and involved in a proliferating cell population (El Cheikh et al. 2014). Mutations in the clock genes CRY1 and CRY2 may lower the growth rate of cells, while PER2 mutations and BMAL1 knockouts may promote it for autonomous stages of the cell cycle shorter than 21 h (El Cheikh et al. 2014). The combination of a molecular model with a population model has been proposed to explain the impacts of the circadian system on the cell population growth (El Cheikh et al. 2014).

Furthermore, alterations such as DNA damage may affect both of the circadian patterns and the cell cycle. The circadian proteins PER1 and TIMELESS (TIM) may interact with the cell cycle checkpoint components including ataxia telangiectasia mutated (ATM)-checkpoint kinase 2 (CHK2) and ataxia telangiectasia and Rad3-related (ATR)-CHK1. Such interactions are critical for the activation of CHK1 and CHK2 in DNA damage (Kondratov and Antoch 2007). In addition, in both normal and stress states, the complex of TIM and TIM-interacting protein (TIPIN) may interact with the DNA replication components to control the DNA replication activities (Kondratov and Antoch 2007).

The circadian regulation of cell cycle may provide the molecular and cellular connections between the rhythmic patterns and problems such as aging and cancer. Factors such as the circadian effects on cell proliferation, metabolism, stress responses, as well as DNA repair may all be involved in the cellular pathogenesis of aging-associated disorders and carcinogenesis (Khapre et al. 2010; also see Fig. 2.1). The circadian genes including PERs, BMAL1, and CRYs may have pivotal roles in such mechanisms.

For example, transgenic mice with mutations in the clock genes have been found to develop cancer and premature aging, as genome integrity and cell proliferation are critical in such disorders (Khapre et al. 2010). In addition, studies have identified circadian variations in the expression of genes associated with genotoxic stress responses (Kondratov and Antoch 2007). These mechanisms may have implications for chronotherapy to promote drug efficacy and tolerance by adjusting the administrative time (see Chap. 7).

2.4 Systems Biology Approaches for Modeling the Circadian Networks

In conclusion, systems biology studies would allow for the integrative description of the individual elements and interactive networks within and among cells toward the understanding of the dynamical principles (Ueda 2007; De Haro and Panda 2006). Both experimental and theoretical modeling methods can be applied to map the interactions at the genome-wide scale to explore the regulatory networks underpinning cellular rhythms.

Systems biology methods such as genetic perturbations and computational modeling within each scale may contribute to the advancement of systems and dynamical medicine (see Chap. 1). Approaches including gene expression profiling and proteomic analyses are re-shaping our understanding of the circadian system. Perturbation analysis and synthetic methods may enhance our knowledge of the mechanisms underlying the transcriptional regulations and robustness of the oscillators (Baggs and Hogenesch 2010).

Such approaches can be applied to identify patterns at various levels including the expression patterns of circadian genes. For example, by analyzing RNA data, promoter elements, and phase information, a circadian transcriptional network was developed with the discovery of the gene-expression patterns around the 24-h clock (Hayes et al. 2005). Factors associated with circadian expression were presented in the network, especially the transcriptional circuits including the REV-ERB/ROR-regulatory element (RRE).

In addition to experimental methods, theoretical frameworks can be applied for studying the dynamical systems using systems theory and nonlinear dynamics. Such frameworks can be useful for modeling the dynamic properties of the "network motifs" and interactions in the molecular and cellular circuits (Zhang et al. 2010b). The network motifs of the feedback and feedforward loops such as those described above are essential for cellular functions including homeostasis and oscillations. The elucidation of the functional networks may help improve the understanding of the complex cellular responses to environmental changes and therapeutics (Zhang et al. 2010b). Such understanding may contribute to the discovery of systemic biomarkers and strategies in chronotherapy (see Chap. 7).

References

Baggs JE, Hogenesch JB (2010) Genomics and systems approaches in the mammalian circadian clock. Curr Opin Genet Dev 20:581–587

Bieler J, Cannavo R, Gustafson K, Gobet C, Gatfield D, Naef F (2014) Robust synchronization of coupled circadian and cell cycle oscillators in single mammalian cells. Mol Syst Biol 10:739

Bussi IL, Levín G, Golombek DA, Agostino PV (2014) Involvement of dopamine signaling in the circadian modulation of interval timing. Eur J Neurosci 40:2299–2310

Cao Q, Gery S, Dashti A, Yin D, Zhou Y, Gu J, Koeffler HP (2009) A role for the clock gene per1 in prostate cancer. Cancer Res 69:7619–7625

Cao R, Robinson B, Xu H, Gkogkas C, Khoutorsky A, Alain T, Yanagiya A, Nevarko T, Liu AC, Amir S et al (2013) Translational control of entrainment and synchrony of the suprachiasmatic circadian clock by mTOR/4E-BP1 signaling. Neuron 79:712–724

Chandra V, Mahajan S, Saini A, Dkhar HK, Nanduri R, Raj EB, Kumar A, Gupta P (2013) Human IL10 gene repression by Rev-erbα ameliorates *Mycobacterium tuberculosis* clearance. J Biol Chem 288:10692–10702

Chong SYC, Ptáček LJ, Fu Y-H (2012) Genetic insights on sleep schedules: this time, it's PERsonal. Trends Genet 28:598–605

Couto P, Miranda D, Vieira R, Vilhena A, De Marco L, Bastos-Rodrigues L (2014) Association between CLOCK, PER3 and CCRN4L with non-small cell lung cancer in Brazilian patients. Mol Med Rep 10:435–440

Dashti HS, Follis JL, Smith CE, Tanaka T, Cade BE, Gottlieb DJ, Hruby A, Jacques PF, Lamon-Fava S, Richardson K et al (2015) Habitual sleep duration is associated with BMI and macronutrient intake and may be modified by CLOCK genetic variants. Am J Clin Nutr 101:135–143

DBBR (2015) The database of biological rhythms. http://pharmtao.com/health/biological-rhythms-database/. Accessed 1 June 2015

De Haro L, Panda S (2006) Systems biology of circadian rhythms: an outlook. J Biol Rhythms 21:507–518

El Cheikh R, Bernard S, El Khatib N (2014) Modeling circadian clock–cell cycle interaction effects on cell population growth rates. J Theor Biol 363:318–331

Fernandes PACM, Cecon E, Markus RP, Ferreira ZS (2006) Effect of TNF-alpha on the melatonin synthetic pathway in the rat pineal gland: basis for a "feedback" of the immune response on circadian timing. J Pineal Res 41:344–350

Fu A, Leaderer D, Zheng T, Hoffman AE, Stevens RG, Zhu Y (2012) Genetic and epigenetic associations of circadian gene TIMELESS and breast cancer risk. Mol Carcinog 51:923–929

Gérard C, Goldbeter A (2012) Entrainment of the mammalian cell cycle by the circadian clock: modeling two coupled cellular rhythms. PLoS Comput Biol 8:e1002516

Goldbeter A, Gérard C, Gonze D, Leloup J-C, Dupont G (2012) Systems biology of cellular rhythms. FEBS Lett 586:2955–2965

Goldsmith CS, Bell-Pedersen D (2013) Diverse roles for MAPK signaling in circadian clocks. Adv Genet 84:1–39

Gotoh T, Vila-Caballer M, Liu J, Schiffhauer S, Finkielstein CV (2015) Association of the circadian factor Period 2 to p53 influences p53's function in DNA-damage signaling. Mol Biol Cell 26:359–372

Hayes KR, Baggs JE, Hogenesch JB (2005) Circadian clocks are seeing the systems biology light. Genome Biol 6:219

Hoffman AE, Zheng T, Ba Y, Stevens RG, Yi C-H, Leaderer D, Zhu Y (2010) Phenotypic effects of the circadian gene Cryptochrome 2 on cancer-related pathways. BMC Cancer 10:110

Hogenesch JB, Ueda HR (2011) Understanding systems-level properties: timely stories from the study of clocks. Nat Rev Genet 12:407–416

Hsuchou H, Wang Y, Cornelissen-Guillaume GG, Kastin AJ, Jang E, Halberg F, Pan W (2013) Diminished leptin signaling can alter circadian rhythm of metabolic activity and feeding. J Appl Physiol 115:995–1003

Hua P, Liu W, Chen D, Zhao Y, Chen L, Zhang N, Wang C, Guo S, Wang L, Xiao H et al (2014) Cry1 and Tef gene polymorphisms are associated with major depressive disorder in the Chinese population. J Affect Disord 157:100–103

Hwang J-W, Sundar IK, Yao H, Sellix MT, Rahman I (2014) Circadian clock function is disrupted by environmental tobacco/cigarette smoke, leading to lung inflammation and injury via a SIRT1-BMAL1 pathway. FASEB J 28:176–194

Khapre RV, Samsa WE, Kondratov RV (2010) Circadian regulation of cell cycle: molecular connections between aging and the circadian clock. Ann Med 42:404–415

Kishi T, Kitajima T, Ikeda M, Yamanouchi Y, Kinoshita Y, Kawashima K, Okochi T, Ozaki N, Iwata N (2008) Association analysis of nuclear receptor Rev-erb alpha gene (NR1D1) with mood disorders in the Japanese population. Neurosci Res 62:211–215

Ko CH, Takahashi JS (2006) Molecular components of the mammalian circadian clock. Hum Mol Genet 15(Spec No 2):R271–R277

Kondratov RV, Antoch MP (2007) Circadian proteins in the regulation of cell cycle and genotoxic stress responses. Trends Cell Biol 17:311–317

Kovanen L, Kaunisto M, Donner K, Saarikoski ST, Partonen T (2013) CRY2 genetic variants associate with dysthymia. PLoS One 8:e71450

Kowalska E, Ripperger JA, Hoegger DC, Bruegger P, Buch T, Birchler T, Mueller A, Albrecht U, Contaldo C, Brown SA (2013) NONO couples the circadian clock to the cell cycle. Proc Natl Acad Sci U S A 110:1592–1599

Lecarpentier Y, Claes V, Duthoit G, Hébert J-L (2014) Circadian rhythms, Wnt/beta-catenin pathway and PPAR alpha/gamma profiles in diseases with primary or secondary cardiac dysfunction. Front Physiol 5:429

Lee JH, Sancar A (2011) Regulation of apoptosis by the circadian clock through NF-kappaB signaling. Proc Natl Acad Sci U S A 108:12036–12041

Lengyel Z, Lovig C, Kommedal S, Keszthelyi R, Szekeres G, Battyáni Z, Csernus V, Nagy AD (2013) Altered expression patterns of clock gene mRNAs and clock proteins in human skin tumors. Tumour Biol 34:811–819

Lewintre EJ, Martín CR, Ballesteros CG, Montaner D, Rivera RF, Mayans JR, García-Conde J (2009) Cryptochrome-1 expression: a new prognostic marker in B-cell chronic lymphocytic leukemia. Haematologica 94:280–284

Lin F, Chen Y, Li X, Zhao Q, Tan Z (2013) Over-expression of circadian clock gene Bmal1 affects proliferation and the canonical Wnt pathway in NIH-3T3 cells. Cell Biochem Funct 31:166–172

Lin C, Tang X, Zhu Z, Liao X, Zhao R, Fu W, Chen B, Jiang J, Qian R, Guo D (2014) The rhythmic expression of clock genes attenuated in human plaque-derived vascular smooth muscle cells. Lipids Health Dis 13:14

Liu C, Li H, Qi L, Loos RJF, Qi Q, Lu L, Gan W, Lin X (2011) Variants in GLIS3 and CRY2 are associated with type 2 diabetes and impaired fasting glucose in Chinese Hans. PLoS One 6:e21464

Luo Y, Wang F, Chen L-A, Chen X-W, Chen Z-J, Liu P-F, Li F-F, Li C-Y, Liang W (2012) Deregulated expression of cry1 and cry2 in human gliomas. Asian Pac J Cancer Prev 13:5725–5728

Maire M, Reichert CF, Gabel V, Viola AU, Strobel W, Krebs J, Landolt HP, Bachmann V, Cajochen C, Schmidt C (2014) Sleep ability mediates individual differences in the vulnerability to sleep loss: evidence from a PER3 polymorphism. Cortex 52:47–59

Masri S, Cervantes M, Sassone-Corsi P (2013) The circadian clock and cell cycle: interconnected biological circuits. Curr Opin Cell Biol 25:730–734

Merino B, Somoza B, Ruiz-Gayo M, Cano V (2008) Circadian rhythm drives the responsiveness of leptin-mediated hypothalamic pathway of cholecystokinin-8. Neurosci Lett 442:165–168

Mullenders J, Fabius AWM, Madiredjo M, Bernards R, Beijersbergen RL (2009) A large scale shRNA barcode screen identifies the circadian clock component ARNTL as putative regulator of the p53 tumor suppressor pathway. PLoS One 4:e4798

Okazaki H, Matsunaga N, Fujioka T, Okazaki F, Akagawa Y, Tsurudome Y, Ono M, Kuwano M, Koyanagi S, Ohdo S (2014) Circadian regulation of mTOR by the ubiquitin pathway in renal cell carcinoma. Cancer Res 74:543–551

Pappa KI, Gazouli M, Anastasiou E, Iliodromiti Z, Antsaklis A, Anagnou NP (2013) The major circadian pacemaker ARNT-like protein-1 (BMAL1) is associated with susceptibility to gestational diabetes mellitus. Diabetes Res Clin Pract 99:151–157

Partonen T (2012) Clock gene variants in mood and anxiety disorders. J Neural Transm 119:1133–1145

Ptitsyn AA, Gimble JM (2007) Analysis of circadian pattern reveals tissue-specific alternative transcription in leptin signaling pathway. BMC Bioinf 8(Suppl 7):S15

Rybakowski JK, Dmitrzak-Weglar M, Kliwicki S, Hauser J (2014) Polymorphism of circadian clock genes and prophylactic lithium response. Bipolar Disord 16:151–158

Sato F, Nagata C, Liu Y, Suzuki T, Kondo J, Morohashi S, Imaizumi T, Kato Y, Kijima H (2009) PERIOD1 is an anti-apoptotic factor in human pancreatic and hepatic cancer cells. J Biochem 146:833–838

Schöning JC, Staiger D (2005) At the pulse of time: protein interactions determine the pace of circadian clocks. FEBS Lett 579:3246–3252

Shimba S, Watabe Y (2009) Crosstalk between the AHR signaling pathway and circadian rhythm. Biochem Pharmacol 77:560–565

Sjöholm LK, Backlund L, Cheteh EH, Ek IR, Frisén L, Schalling M, Osby U, Lavebratt C, Nikamo P (2010) CRY2 is associated with rapid cycling in bipolar disorder patients. PLoS One 5:e12632

Štorcelová M, Vicián M, Reis R, Zeman M, Herichová I (2013) Expression of cell cycle regulatory factors hus1, gadd45a, rb1, cdkn2a and mre11a correlates with expression of clock gene per2 in human colorectal carcinoma tissue. Mol Biol Rep 40:6351–6361

Truong T, Liquet B, Menegaux F, Plancoulaine S, Laurent-Puig P, Mulot C, Cordina-Duverger E, Sanchez M, Arveux P, Kerbrat P et al (2014) Breast cancer risk, nightwork, and circadian clock gene polymorphisms. Endocr Relat Cancer 21:629–638

Ueda HR (2007) Systems biology of mammalian circadian clocks. Cold Spring Harb Symp Quant Biol 72:365–380

Ukai H, Ueda HR (2010) Systems biology of mammalian circadian clocks. Annu Rev Physiol 72:579–603

Vanderlind WM, Beevers CG, Sherman SM, Trujillo LT, McGeary JE, Matthews MD, Maddox WT, Schnyer DM (2014) Sleep and sadness: exploring the relation among sleep, cognitive control, and depressive symptoms in young adults. Sleep Med 15:144–149

Wallach T, Schellenberg K, Maier B, Kalathur RKR, Porras P, Wanker EE, Futschik ME, Kramer A (2013) Dynamic circadian protein–protein interaction networks predict temporal organization of cellular functions. PLoS Genet 9:e1003398

Wang X, Yu W, Zheng L (2015) The dynamics of NF-κB pathway regulated by circadian clock. Math Biosci 260:47–53

Weigl Y, Ashkenazi IE, Peleg L (2013) Rhythmic profiles of cell cycle and circadian clock gene transcripts in mice: a possible association between two periodic systems. J Exp Biol 216:2276–2282

Yamada Y, Forger D (2010) Multiscale complexity in the mammalian circadian clock. Curr Opin Genet Dev 20:626–633

Yang G, Wright CJ, Hinson MD, Fernando AP, Sengupta S, Biswas C, La P, Dennery PA (2014) Oxidative stress and inflammation modulate Rev-erbα signaling in the neonatal lung and affect circadian rhythmicity. Antioxid Redox Signal 21:17–32

Yasuniwa Y, Izumi H, Wang K-Y, Shimajiri S, Sasaguri Y, Kawai K, Kasai H, Shimada T, Miyake K, Kashiwagi E et al (2010) Circadian disruption accelerates tumor growth and angio/stromagenesis through a Wnt signaling pathway. PLoS One 5:e15330

Yoshida K, Sato M, Hase T, Elshazley M, Yamashita R, Usami N, Taniguchi T, Yokoi K, Nakamura S, Kondo M et al (2013) TIMELESS is overexpressed in lung cancer and its expression correlates with poor patient survival. Cancer Sci 104:171–177

Yu H, Meng X, Wu J, Pan C, Ying X, Zhou Y, Liu R, Huang W (2013) Cryptochrome 1 overexpression correlates with tumor progression and poor prognosis in patients with colorectal cancer. PLoS One 8:e61679

Zeng Z, Luo H, Yang J, Wu W, Chen D, Huang P, Xu R (2014) Overexpression of the circadian clock gene Bmal1 increases sensitivity to oxaliplatin in colorectal cancer. Clin Cancer Res 20:1042–1052

Zhang EE, Kay SA (2010) Clocks not winding down: unravelling circadian networks. Nat Rev Mol Cell Biol 11:764–776

Zhang EE, Liu Y, Dentin R, Pongsawakul PY, Liu AC, Hirota T, Nusinow DA, Sun X, Landais S, Kodama Y et al (2010a) Cryptochrome mediates circadian regulation of cAMP signaling and hepatic gluconeogenesis. Nat Med 16:1152–1156

Zhang Q, Bhattacharya S, Andersen ME, Conolly RB (2010b) Computational systems biology and dose–response modeling in relation to new directions in toxicity testing. J Toxicol Environ Health B Crit Rev 13:253–276

Zhao N, Yang K, Yang G, Chen D, Tang H, Zhao D, Zhao C (2013) Aberrant expression of clock gene period1 and its correlations with the growth, proliferation and metastasis of buccal squamous cell carcinoma. PLoS One 8:e55894

Chapter 3
The Circadian–Immune Crosstalk and Inflammation: Implications for Disease Treatment

3.1 A Systemic Overview of the Bidirectional Circadian–Immune Interactions

Chronic inflammation is an important risk factor for many diseases including cardiovascular disease and cancer. Chronic and systemic inflammatory conditions are characterized with the entrance of inflammatory macrophages into various tissues. Cellular transformation has been closely related to inflammation via factors such as cyclooxygenase-2 (COX-2) and reactive oxygen intermediate (ROI) (Baldassarre et al. 2004). The malfunction of cyclooxygenase and nitric oxide synthase may affect the expression of proteins controlling the cell cycle progression. Complex cellular pathways may be involved in the inflammatory responses and abnormal cell cycle progression.

Circadian disruptions may have an essential role in inflammation. Significant daily rhythms have been detected among various immune factors including the monocyte chemotactic protein, granulocyte-macrophage colony-stimulating factor, interleukin 8 (IL8), and tumor necrosis factor α (TNFα) (Rahman et al. 2014). The disturbance of circadian rhythms may lead to a broad range of pathophysiological and clinical states, from inflammation to type 2 diabetes (Sato et al. 2014).

Figure 3.1 provides a schematic overview of the interactions between the circadian clocks and immune mediators at various levels in association with inflammatory responses. The environmental factors such as light serve as the external cues that may entrain the circadian rhythms. The hypothalamic suprachiasmatic nucleus (SCN) hosts the central clock controlling the synchronization of the rhythms. Working together with the hypothalamic–pituitary–adrenal (HPA) axis, the humoral and neural output systems regulate the oscillations in the peripheral tissues including the spleen, liver, and bone marrows (see Fig. 3.1).

At the system level, the HPA axis has a critical role in the complex circadian–immune interactions. Hormones such as melatonin from the pineal gland and glucocorticoids from the adrenal gland participate in the regulation of the rhythmic patterns (see Fig. 3.1). At the molecular and cellular levels, the circadian and

© Springer International Publishing Switzerland 2015   25
Q. Yan, *Cellular Rhythms and Networks: Implications for Systems Medicine*,
SpringerBriefs in Cell Biology, DOI 10.1007/978-3-319-22819-8_3

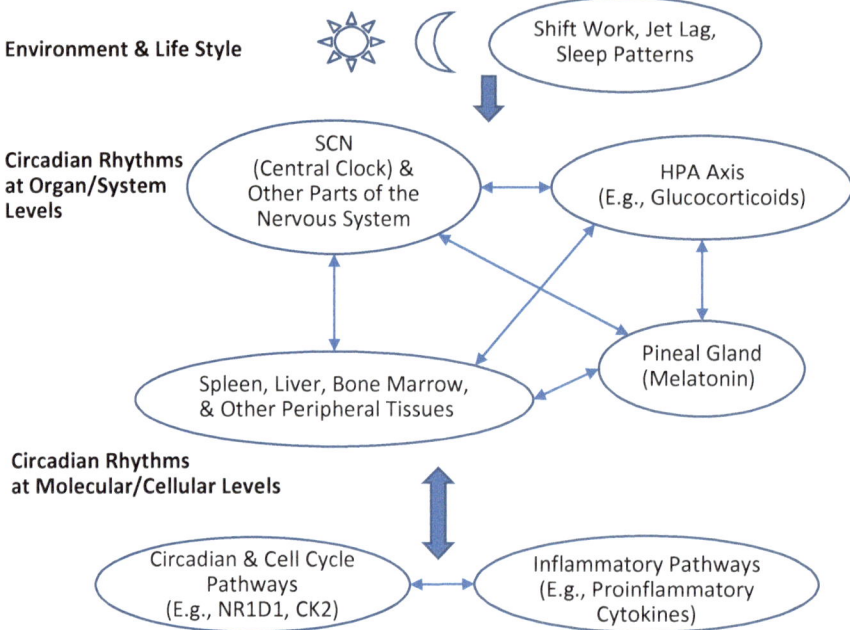

Fig. 3.1 The circadian–immune interactions at various levels in association with inflammation

immune pathways have reciprocal impacts on each other. For example, the clock proteins including REV-ERBα (also known as NR1D1) may have a negative control on the cytokine gene expressions. On the other hand, proinflammatory cytokines may convey feedback to the circadian molecular clocks and cell cycles (Arjona et al. 2012; also see Fig. 3.1). In another example, the highly conserved serine-threonine kinase casein kinase 2 (CK2) has a key role in the regulation of various activities including circadian rhythms, cell cycle progression, apoptosis, and inflammatory responses (Singh and Ramji 2008).

3.1.1 The Circadian–Immune Interactions at the Molecular and Cellular Levels

In accordance with the environmental cycle, the circadian clock arranges the temporal patterns of physiology and psychological activities via the integration of the rhythms in cells and tissues all over the body. Physiologically, the circadian rhythms are vital for normal immune functions and responses. The entrainment of the clockwork in the SCN may supply essential signals for the robust rhythms in the immune cells for maintaining processes including immune-surveillance and memory T-cell activities (Prendergast et al. 2013).

As mentioned earlier, the interconnections between the immune system and the circadian oscillating systems are bidirectional at multiple levels. The immune system is controlled by circadian rhythms in various aspects including the quantity of red blood cells and peripheral blood mononuclear cells (Cermakian et al. 2014). The concentrations of essential immune mediators such as cytokines have daily oscillation patterns. Furthermore, the crosstalk between the circadian system and immune tissue cells involves the daily patterns of autonomic and endocrine oscillations. The cytokine levels may also have impacts on the clock, indicating the complex communication of circadian information between the neuroendocrine and immune cells (Cermakian et al. 2014).

In addition, the relationships between inflammation and genetic alterations in the regulation of cell cycle checkpoints may be crucial for decreasing the malignant transformation of the inflammatory disorders (Baldassarre et al. 2004). Different endogenous and extrinsic factors may be involved in inflammatory responses and cell cycle alterations. In human cumulus granulosa cells, adiposity has been related to the abnormal expressions of the genes associated with cell cycle progression and inflammation. These changes include the up-regulation of protein phosphatase 1-like (PPM1L) and down-regulation of cell division cycle 20 (CDC20) (Merhi et al. 2015). In another example, cigarette smoke may lead to the activation of pro-inflammatory pathways, the altered expressions of p53 and p21, and the abnormal progression of the cell cycle in lung fibroblasts (D'Anna et al. 2015). Furthermore, the high mobility group A1 (HMGA1) protein may activate inflammatory pathways during the early stages of lymphoid tumorigenesis, and the pathways associated with cell cycle progression in established tumors (Schuldenfrei et al. 2011).

These observations indicate that various molecules and pathways are involved in the circadian–immune interactions. For example, in response to inflammatory signals, USP2a, the circadian-regulated deubiquitinating enzyme, may interact with the clock protein cryptochrome1 (CRY1) and promote its stability (Tong et al. 2012). In addition, the proinflammatory cytokine TNFα may also interact with the CRY1 protein in an USP2a-associated manner. More examples will be given in the next section.

3.1.2 Examples of Molecular Factors and Cellular Networks

Table 3.1 lists some examples of circadian-associated molecular factors and cellular networks that may have important roles in inflammation and relevant diseases. Some of these factors are discussed in details below. A more complete list can be found in the Database of Biological Rhythms (DBBR 2015).

3.1.2.1 CRY1 and Associated Networks

The circadian-oscillator components such as CRY1 have essential roles in the regulation of immune responses. CRY1 may be involved in NFκB and TNFα pathways associated with chronic inflammation. Lower levels of CRY proteins may lead to

Table 3.1 Examples of circadian-associated genes and pathways involved in inflammation

Genes	Changes/interactions	Associated disorders	References
CRY1	Altered melatonin, lower CRY1 and PER2, higher TNFα and IL6	Early-stage sepsis	Li et al. (2013)
	Lower CRY1, higher TNFα, IL1β, IL-6, WEE1, c-Fos	Rheumatoid arthritis (RA)	Hashiramoto et al. (2010)
	Lower CRY1, higher cAMP, PKA, NFκB	Chronic inflammatory diseases	Narasimamurthy et al. (2012)
	Overexpression, NFκB and cAMP/PKA pathways	Therapy of vascular inflammation	Qin and Deng (2015)
IL6	Rhythms of TNFα, IL1β, IL6	Microglia inflammation	Fonken et al. (2015)
	Altered IL6, STAT3, KCNV2, CAMK2D	Altered sleep homeostasis, circadian disruptions	Möller-Levet et al. (2013)
	Higher IL6	RA morning symptoms	Alten (2012)
	Altered cortisol and IL6	Circadian variation in RA	Perry et al. (2009)
	Higher IL6, lower glucocorticoid sensitivity	Fibromyalgia	Geiss et al. (2012)
	Higher coronary levels of IL6 in the afternoon	Morbidity in myocardial infarction	Bonda et al. (2010)
SIRT1	SIRT1-BMAL1 pathway in circadian disruptions	Lung injury from tobacco/cigarette smoke	Hwang et al. (2014)
	Interaction with RelB	Acute inflammation, altered mitochondrial bioenergetics	Millet et al. (2013)
	Altered SIRT1 in energy metabolism and circadian	Obesity-induced inflammation	Li (2013)
	PARP1, NFκB	Chronic inflammation, aging, cigarette smoke	Hwang et al. (2013)
	AMPK, adiponectin, circadian clock in metabolic tissues	Metabolic diseases associated with obesity	Schug and Li (2011)
	Adiponectin, circadian rhythms	Type 2 diabetes	Kitada and Koya (2013) and Kitada et al. (2013)
	Metabolism, circadian rhythms	Cancer	Bruzzone et al. (2013)
TNF	IL6, HPA axis dysfunction	Chronic inflammation, metabolic syndrome	Straub et al. (2011)
	Higher TNFα, phase shifts in IL6, HPA axis dysfunction	Collagen-induced arthritis (CIA)	Li et al. (2004)
	PER2, DBP, HLF, TEF, E4BP4	RA	Yoshida et al. (2013)
	Higher BMI, stress	Cardiovascular events	Wirtz et al. (2008)
	Adenosine A2A receptor, circadian in immune cells	RA	Perez-Aso et al. (2013)
	Higher TNFα, VEGF, IL6, PER3 variants	Poor sleep, fatigue, cancer	Guess et al. (2009)

lower inhibition on cAMP and higher levels of PKA activation (Narasimamurthy et al. 2012). Such changes may result in the phosphorylation of p65 at S276 and NFκB activation. On the other hand, higher levels of CRY1 expression may suppress sleep deprivation-caused vascular inflammation associated with the NFκB and cAMP/PKA pathways (Qin and Deng 2015).

In the early stages of sepsis, altered melatonin production and lower expressions of CRY1 and PER2 have been observed together with higher expressions of proinflammatory cytokines TNFα and IL6 (Li et al. 2013). PER2 has been found as an essential regulator of the functions of NK cells, serving as the direct connection between the circadian systems and innate immune responses (Liu et al. 2006).

The interactions between CRY1 and TNFα have been associated with various diseases such as rheumatoid arthritis (Hashiramoto et al. 2010). Together with other circadian genes including CLOCK, PER3, and RORA, the genetic alterations in peripheral blood leukocytes represent a feature of acute systemic inflammation that is uncoupled from the activity of the central clock (Haimovich et al. 2010).

Moreover, factors such as chronic sleep deprivation may disturb the circadian pathways involving CRY1 and lead to higher susceptibility to chronic inflammatory diseases including diabetes, obesity, and cancer (Narasimamurthy et al. 2012). As higher inflammation is the common link among these chronic diseases, the key roles of the circadian genes and pathways make them the potential therapeutic targets.

3.1.2.2   SIRT1 and Associated Networks

Sirtuin 1 (SIRT1) is involved in a wide range of cellular physiological and pathological processes via the deacetylation of transcription factors and histones. These processes include mitochondrial biogenesis, cellular survival and senescence, apoptosis, endothelial functions, inflammation, as well as circadian rhythms (Hwang et al. 2013; Yamakuchi 2012). As a metabolic sensor in various tissues, SIRT1 provides the linkage among the cellular metabolism, gene expression regulation, and stress responses (Li 2013). Involving in the circadian oscillations in peripheral tissues, it may serve as the molecular, cellular, and functional connection among energy homeostasis, chromatin modification, and circadian physiology (Bellet and Sassone-Corsi 2010).

For example, in oxidative stress-associated chronic inflammation and aging, lower levels of SIRT1 have been observed. Such decrease may disturb its regulation of the acetylation of target proteins including p53, RELA/p65, and FOXO3 (Hwang et al. 2013). These changes may promote the inflammatory and stress responses, as well as the processes of cellular senescence and endothelial dysfunction.

Furthermore, SIRT1 has a key role in metabolic pathways associated with diseases such as diabetes, cardiovascular diseases, and cancer (Kitada and Koya 2013). It has been suggested as the therapeutic target for type 2 diabetes and cancer (Kitada et al. 2013). It is involved in the regulation of glucose and lipid metabolism in the liver, fat mobilization in white adipose tissue, nutrient availability senses in the hypothalamus, insulin production in the pancreas, and obesity-associated inflammation in macrophages (Li 2013). In addition, environmental factors such as tobacco/cigarette smoke may induce circadian disruptions via the SIRT1-BMAL1 pathway, leading to lung inflammation and injury in the obstructive lung diseases (Hwang et al. 2014).

3.1.3   The Circadian–Immune Crosstalk at the Tissue and Organism Levels

The mechanisms of the circadian–immune crosstalk suggest potential strategies in reprogramming biological rhythms and adjusting homeostatic oscillations to promote therapeutic outcomes (Mavroudis et al. 2013). The design of better treatment methods require investigations at various levels from molecular factors to cellular pathways, from alterations in various tissues to environmental interactions.

Systemic studies have shown that internal circadian clockworks are running autonomously in the spleen, lymph nodes, and peritoneal macrophages (see Fig. 3.1). These clocks control the rhythms associated with the inflammatory responses. For instance, in isolated spleen cells, the stimulation of bacterial endotoxin at different time-of-day was associated with the circadian rhythms in TNFα and IL6 productions (Keller et al. 2009). Such oscillations may be driven by a local circadian clock in splenic macrophages. Some parts of the macrophage transcriptome such as the regulators for pathogen detection and cytokine production may oscillate in a circadian manner (Keller et al. 2009).

The circadian-associated factors can be affected by immune challenges at different temporal, genetic, and tissue-specific scales, as well as by body temperature at the organismal level (Cermakian et al. 2014). A broad spectrum of rhythmically regulated humoral elements may be associated with the inflammatory pathways, including glucocorticoids, prostaglandins, melatonin, and leptin. Other factors also have critical roles, such as the neuronal associations between the brain and peripheral tissues, as well as the rhythmic activities of cytokines and their receptors (Cermakian et al. 2014).

Complex networks and feedback loops among the circadian–neuroendocrine–immune interrelationships are involved in the immune responses in different tissues and organs. However, such endogenous synchronization can often be disturbed by environmental factors such as chronic stress. For example, chronically stressed patients often show diminished rhythmic properties of endocrine signals (Cermakian et al. 2014). A systemic understanding incorporating psychological and environmental interactions is necessary to achieve the integrative view of the complex crosstalk.

3.2   The Circadian–Inflammation Associations in Diseases

As discussed above, the circadian–immune crosstalk is critical for understanding the temporal factors in pathophysiological immune responses. Circadian disruptions may have critical roles in the development and progression of many diseases including inflammatory, metabolic, and alcohol-related disorders. For example, circadian misalignment alone even without sleep loss (such as the condition in shift work) has been found to elevate inflammatory markers and insulin resistance with higher risks for diabetes (Leproult et al. 2014). The multi-factorial interactions among circadian patterns, immune responses, metabolism processes, and epigenetic activities may be involved in these processes (Voigt et al. 2013). In the following sections, the circadian–immune crosstalk and systemic inflammatory factors will be discussed in various diseases.

3.2.1   Lung Disorders

Patients with lung disorders such as obstructive lung diseases often have abnormal circadian patterns in the lung functions (Hwang et al. 2014). As the immune functions are regulated by circadian rhythms, inflammatory lung diseases often show circadian alterations in symptom seriousness. Local and systemic circadian clocks may interact with each other in both physiological and pathological conditions. As indicated in a study using a mice model, a pulmonary epithelial cell clock is involved in the regulation of neutrophil gathering to the lungs in association with the lung inflammation (Thompson et al. 2014).

At the molecular and cellular levels, systemic inflammation may alter the expression of clock genes. For example, a genome-wide study using a mice model with endotoxin treatment showed that the circadian patterns of leukocyte counts in the lung were changed in a BMAL1-dependent way (Haspel et al. 2014). In addition, the granulocytes rather than lymphocytes became the major oscillating cell type. Such analysis indicates that the cellular circadian rhythms may be changed and re-organized by the inflammatory processes in association with the lung injury.

The interactions between cytokines and circadian genes may have crucial roles in the lung inflammation with complex immune-endocrine mechanisms involved. In epithelial club (Clara) cells, the circadian clocks are involved in pulmonary antibacterial responses, while the chemokine CXCL5 is associated with the circadian neutrophil recruitment to the lung (Gibbs et al. 2014). Among the bronchiolar cells, the disruption of rhythmic CXCL5 expression may be caused by alterations in the clock gene BMAL1. Such changes may lead to higher inflammatory responses to lipopolysaccharide and compromised host reactions to the infections caused by Streptococcus pneumoniae. Furthermore, the adrenal axis and glucocorticoid hormones are involved in the time-of-day alterations, the responses to bacterial infections, and the pulmonary inflammation (Gibbs et al. 2014).

In another example, the transcription of the circadian gene REV-ERBα (also known as NR1D1) may be affected by oxidative stress and inflammation in the mouse neonatal lung (Yang et al. 2014). Because of the involvement of REV-ERBα in cellular metabolism and its interaction with NFκB, such mechanisms have been associated with the cellular dysfunction of the lung and lung injury.

As mentioned earlier, the gene–environment interactions are also pivotal. Environmental tobacco/cigarette smoke (CS) may disrupt the circadian clocks mediated by the SIRT1-BMAL1 pathway and result in lung injury. In the mice exposed to CS and among patients with chronic obstructive pulmonary disease (COPD), altered expressions of SIRT1 and BMAL1 were observed in the lung epithelial cells with higher inflammatory levels (Hwang et al. 2014). In another example, sub-chronic secondhand tobacco smoke exposure caused higher levels of lung neutrophils and pulmonary CYP1A1, indicating higher levels of oxidative stress and lung inflammation (Gentner and Weber 2012).

3.2.2   Cardiovascular Diseases

The elucidation of the kinetics of inflammation and the diurnal variations may contribute to the discovery and validation of useful biomarkers for cardiovascular diseases. Systemic inflammation has been closely related to the alterations of cardiac autonomic modulation (CAM) and risks for cardiac disease. Studies using middle-aged samples found that systemic inflammation may be adversely associated with the circadian patterns of CAM (Li et al. 2011). Circadian disruptions have been linked to arterial inflammation and the occurrence of cardiovascular accidents including acute coronary syndrome (Dominguez-Rodriguez et al. 2011). More discussions on the circadian-cardiovascular interactions can be found in Chap. 5. This section will focus on inflammation as the important connection.

For example, as a key element of the circadian system, REV-ERBα (also known as NR1D1) has key roles in the regulation of metabolism, blood pressure, cardiac homeostasis, and inflammation (Ramakrishnan and Muscat 2006). The REV-ERB (NR1D) subgroup including REV-ERBα/NR1D1 and REV-ERBβ/NR1D2 are abundant in peripheral tissues with heavy energy consumptions including the skeletal muscle, brown and white adipose, liver, kidney, and brain. In vascular smooth and skeletal muscle cells, the NR1D subgroup is involved in the inflammatory responses via the regulation of NFκB associated pathways (Ramakrishnan and Muscat 2006).

Furthermore, lifestyle and environmental factors may also have critical roles. For instance, circadian disruptions and higher risks for metabolic, hepatic, cardiovascular and neurodegenerative disorders have been reported among shift workers (Summa et al. 2013). These conditions may be caused by the pathologic inflammatory states. In addition, changes caused by tobacco smoke have been associated with not only arterial stiffness, but also altered circadian patterns of both heart rate and blood pressure. Tobacco smoke exposure may be a risk factor for cardiovascular diseases with inflammatory damages in both of the structural and functional properties in the arteries (Gentner and Weber 2012).

3.2.3 Gastrointestinal Disorders

The physiological functions of the gastrointestinal tract are also controlled by the circadian clocks. Circadian disruptions may elevate the permeability of the intestinal epithelial barriers including gut leakiness (Summa et al. 2013). Such changes may lead to the penetration of proinflammatory bacterial materials (such as endotoxin) through the intestinal wall to enter the systemic circulation.

For example, studies using a model of chronic alcohol consumption observed that genetic and environmental circadian disruptions were associated with intestinal hyperpermeability (Summa et al. 2013). Such mechanisms are involved in the higher levels of alcohol-caused gut leakiness as well as endotoxemia and fatty liver disease. At the molecular level, such states have been related to the tight junction protein occluding (Summa et al. 2013).

These findings suggest that circadian integration is essential for the health of the gastrointestinal tract. On the other hand, circadian disruptions have been considered an important risk factor for alcoholic liver disease, intestinal hyperpermeability, and other illnesses involving endotoxin-associated inflammatory conditions (Summa et al. 2013).

3.2.4 Aging, Arthritis, Cancer, and Other Complex Diseases

The circadian–immune crosstalk may have essential roles in aging, cancer, and other complex diseases. As mentioned earlier, one of the key mediators in such mechanisms is REV-ERBα (also known as NR1D1) (Sato et al. 2014). Specifically, REV-ERBα (NR1D1) may be associated with the regulation of inflammatory activities of macrophages via the inhibition of the CCL2 expression. Many elements may be involved in this process, including a REV-ERBα-binding motif in the CCL2 promoter, the inhibition of cell adhesion and migration in the inflammatory responses, and the ERK- and p38-signaling pathways (Sato et al. 2014). Because it may facilitate the circadian control of innate immunity via the modulation of cytokines, REV-ERBα (NR1D1) has been suggested as a therapeutic target for the treatment of inflammation-associated diseases (Gibbs et al. 2012).

The interactions between the clock genes and inflammatory cytokine pathways have crucial roles in the pathogenesis of rheumatoid arthritis (RA). For instance, joint stiffness is one of the common complaints of RA that often reaches the most serious point in the morning. In the meantime, the secretion of proinflammatory cytokines such as IL6 also reaches the highest level early in the morning, serving as the main factors contributing to the morning stiffness (Yoshida et al. 2014). The understanding of such mechanisms is useful for strategies in chronotherapy (see Chap. 7).

In cancer patients, the cancer cachexia syndrome with involuntary weight loss has been associated with the loss of skeletal muscle and adipose tissues, as well as the higher levels of acute phase proteins and inflammatory cytokines including IL6 (Tsoli et al. 2014). Concurrent disruptions in circadian rhythms and lipid metabolism associated networks such as the AMPK/mTOR pathways may have major roles in these processes (Tsoli et al. 2014).

Moreover, lifestyle is a pivotal factor in the circadian–immune associated complex diseases. For instance, total sleep deprivation may lead to chronic circadian misalignment, affecting the levels of cortisol and pro- and anti-inflammatory proteins. Specifically, circadian misalignment may lead to the elevation in the plasma levels of TNFα and C-reactive protein (CRP) (Wright et al. 2015). These factors have been considered as potential biomarkers for many diseases and treatment responses, as well as for chronotherapy.

3.3 Potential Strategies in Chronotherapy for Inflammation

In conclusion, together with environmental and behavioral cycles, the internal circadian systems may coordinate the temporal constitutions of immune responses (Rahman et al. 2014). The alterations in these complex processes such as circadian disruptions and sleep misalignment may have profound effects on immune functions with implications for disease treatment. These mechanisms indicate that disease predisposition and therapeutic effectiveness may also have such patterns in accordance with the time-of-day (Rahman et al. 2014).

For example, as the primary immune cell type in the central nervous system, microglia has a critical role in the diurnal variations during sickness responses involving inflammatory pathways. Experiments using rat models showed that hippocampal microglia have robust rhythmical expressions of inflammatory factors and clock genes including TNFα, IL1β, and IL6 (Fonken et al. 2015). Peak cytokine gene expressions were observed during the middle of the light phase. These findings suggest that the time-of-day is a key factor in inflammatory interventions, especially when planning for immunotherapies or surgeries (Fonken et al. 2015).

In the example of multiple sclerosis (MS), inflammatory serum parameters have been considered important biomarkers for the indication of disease conditions and therapeutic responses. However, the measurements often have contradictory results because of the diurnal changes in the serum concentrations of the inflammatory biomarkers. Such mechanism makes the blood collection time an important factor although it is often neglected (Wipfler et al. 2013). Specifically, elevated levels of soluble TNF-Receptor-1 (TNF-R1) and TNF-R2 are normally observed in the morning, with lower levels of soluble intercellular adhesion molecule-1 (ICAM-1) and soluble vascular adhesion molecule-1 (sVCAM-1) in the afternoon (Wipfler et al. 2013). However among MS patients with active disease, elevated serum levels of VCAM-1 were observed around noon and in the early afternoon (Wipfler et al. 2013).

Furthermore, some of the integrative interventions may have their effects mediated through the circadian regulation of inflammation. For example, long term restricted feeding (RF) that limits the duration of food availability without calorie restriction has been found to entrain the circadian clocks in peripheral tissues (Sherman et al. 2011). A study based on mouse models evaluated the circadian expressions of clock genes as well as metabolic, inflammatory, and allergy biomarkers. The study found that under RF, the circadian rhythmicity became more robust with lower levels of inflammatory biomarkers including IL6, TNFα, and NFκB (Sherman et al. 2011). Meanwhile, higher levels of the anti-inflammatory cytokine IL10 were detected in the liver and jejunum.

In another example, a study using a rat model observed that melatonin may have robust chronobiotic effects via the promotion of circadian synchronization in animals with chronic inflammation (Laste et al. 2013). These studies indicate that the interventions such as RF and melatonin may have beneficial effects via their impacts on the circadian–inflammation interactions. A more detailed discussion of chronotherapy for various diseases will be provided in Chap. 7.

References

Alten R (2012) Chronotherapy with modified-release prednisone in patients with rheumatoid arthritis. Expert Rev Clin Immunol 8:123–133

Arjona A, Silver AC, Walker WE, Fikrig E (2012) Immunity's fourth dimension: approaching the circadian–immune connection. Trends Immunol 33:607–612

Baldassarre G, Nicoloso MS, Schiappacassi M, Chimienti E, Belletti B (2004) Linking inflammation to cell cycle progression. Curr Pharm Des 10:1653–1666

Bellet MM, Sassone-Corsi P (2010) Mammalian circadian clock and metabolism – the epigenetic link. J Cell Sci 123:3837–3848

Bonda T, Kaminski KA, Kozuch M, Kozieradzka A, Wojtkowska I, Dobrzycki S, Kralisz P, Nowak K, Prokopczuk P, Musial WJ (2010) Circadian variations of interleukin 6 in coronary circulations of patients with myocardial infarction. Cytokine 50:204–209

Bruzzone S, Parenti MD, Grozio A, Ballestrero A, Bauer I, Del Rio A, Nencioni A (2013) Rejuvenating sirtuins: the rise of a new family of cancer drug targets. Curr Pharm Des 19:614–623

Cermakian N, Westfall S, Kiessling S (2014) Circadian clocks and inflammation: reciprocal regulation and shared mediators. Arch Immunol Ther Exp (Warsz) 62:303–318

D'Anna C, Cigna D, Costanzo G, Ferraro M, Siena L, Vitulo P, Gjomarkaj M, Pace E (2015) Cigarette smoke alters cell cycle and induces inflammation in lung fibroblasts. Life Sci 126:10–18

DBBR (2015) The database of biological rhythms. http://pharmtao.com/health/biological-rhythms-database/. Accessed 1 June 2015

Dominguez-Rodriguez A, Tome MC-P, Abreu-Gonzalez P (2011) Interrelation between arterial inflammation in acute coronary syndrome and circadian variation. World J Cardiol 3:57–58

Fonken LK, Frank MG, Kitt MM, Barrientos RM, Watkins LR, Maier SF (2015) Microglia inflammatory responses are controlled by an intrinsic circadian clock. Brain Behav Immun 45:171–179

Geiss A, Rohleder N, Anton F (2012) Evidence for an association between an enhanced reactivity of interleukin-6 levels and reduced glucocorticoid sensitivity in patients with fibromyalgia. Psychoneuroendocrinology 37:671–684

Gentner NJ, Weber LP (2012) Secondhand tobacco smoke, arterial stiffness, and altered circadian blood pressure patterns are associated with lung inflammation and oxidative stress in rats. Am J Physiol Heart Circ Physiol 302:H818–H825

Gibbs J, Ince L, Matthews L, Mei J, Bell T, Yang N, Saer B, Begley N, Poolman T, Pariollaud M et al (2014) An epithelial circadian clock controls pulmonary inflammation and glucocorticoid action. Nat Med 20:919–926

Gibbs JE, Blaikley J, Beesley S, Matthews L, Simpson KD, Boyce SH, Farrow SN, Else KJ, Singh D, Ray DW et al (2012) The nuclear receptor REV-ERBα mediates circadian regulation of innate immunity through selective regulation of inflammatory cytokines. Proc Natl Acad Sci U S A 109:582–587

Guess J, Burch JB, Ogoussan K, Armstead CA, Zhang H, Wagner S, Hebert JR, Wood P, Youngstedt SD, Hofseth LJ et al (2009) Circadian disruption, Per3, and human cytokine secretion. Integr Cancer Ther 8:329–336

Haimovich B, Calvano J, Haimovich AD, Calvano SE, Coyle SM, Lowry SF (2010) In vivo endotoxin synchronizes and suppresses clock gene expression in human peripheral blood leukocytes. Crit Care Med 38:751–758

Hashiramoto A, Yamane T, Tsumiyama K, Yoshida K, Komai K, Yamada H, Yamazaki F, Doi M, Okamura H, Shiozawa S (2010) Mammalian clock gene cryptochrome regulates arthritis via proinflammatory cytokine TNF-alpha. J Immunol 184:1560–1565

Haspel JA, Chettimada S, Shaik RS, Chu J-H, Raby BA, Cernadas M, Carey V, Process V, Hunninghake GM, Ifedigbo E et al (2014) Circadian rhythm reprogramming during lung inflammation. Nat Commun 5:4753

Hwang J, Yao H, Caito S, Sundar IK, Rahman I (2013) Redox regulation of SIRT1 in inflammation and cellular senescence. Free Radic Biol Med 61:95–110

Hwang J-W, Sundar IK, Yao H, Sellix MT, Rahman I (2014) Circadian clock function is disrupted by environmental tobacco/cigarette smoke, leading to lung inflammation and injury via a SIRT1-BMAL1 pathway. FASEB J 28:176–194

Keller M, Mazuch J, Abraham U, Eom GD, Herzog ED, Volk H-D, Kramer A, Maier B (2009) A circadian clock in macrophages controls inflammatory immune responses. Proc Natl Acad Sci U S A 106:21407–21412

Kitada M, Koya D (2013) SIRT1 in type 2 diabetes: mechanisms and therapeutic potential. Diabetes Metab J 37:315–325

Kitada M, Kume S, Kanasaki K, Takeda-Watanabe A, Koya D (2013) Sirtuins as possible drug targets in type 2 diabetes. Curr Drug Targets 14:622–636

Laste G, Vidor L, de Macedo IC, Rozisky JR, Medeiros L, de Souza A, Meurer L, de Souza ICC, Torres ILS, Caumo W (2013) Melatonin treatment entrains the rest-activity circadian rhythm in rats with chronic inflammation. Chronobiol Int 30:1077–1088

Leproult R, Holmbäck U, Van Cauter E (2014) Circadian misalignment augments markers of insulin resistance and inflammation, independently of sleep loss. Diabetes 63:1860–1869

Li X (2013) SIRT1 and energy metabolism. Acta Biochim Biophys Sin (Shanghai) 45:51–60

Li C-X, Liang D-D, Xie G-H, Cheng B-L, Chen Q-X, Wu S-J, Wang J-L, Cho W, Fang X-M (2013) Altered melatonin secretion and circadian gene expression with increased proinflammatory cytokine expression in early-stage sepsis patients. Mol Med Rep 7:1117–1122

Li S, Lu A, Li B, Wang Y (2004) Circadian rhythms on hypothalamic–pituitary–adrenal axis hormones and cytokines of collagen induced arthritis in rats. J Autoimmun 22:277–285

Li X, Shaffer ML, Rodríguez-Colón SM, He F, Bixler EO, Vgontzas AN, Wolbrette DL, Wu C, Ball RW, Liao D (2011) Systemic inflammation and circadian rhythm of cardiac autonomic modulation. Auton Neurosci 162:72–76

Liu J, Malkani G, Mankani G, Shi X, Meyer M, Cunningham-Runddles S, Ma X, Sun ZS (2006) The circadian clock Period 2 gene regulates gamma interferon production of NK cells in host response to lipopolysaccharide-induced endotoxic shock. Infect Immun 74:4750–4756

Mavroudis PD, Scheff JD, Calvano SE, Androulakis IP (2013) Systems biology of circadian–immune interactions. J Innate Immun 5:153–162

Merhi Z, Polotsky AJ, Bradford AP, Buyuk E, Chosich J, Phang T, Jindal S, Santoro N (2015) Adiposity alters genes important in inflammation and cell cycle division in human cumulus granulosa cell. Reprod Sci pii:1933719115572484

Millet P, McCall C, Yoza B (2013) RelB: an outlier in leukocyte biology. J Leukoc Biol 94:941–951

Möller-Levet CS, Archer SN, Bucca G, Laing EE, Slak A, Kabiljo R, Lo JCY, Santhi N, von Schantz M, Smith CP et al (2013) Effects of insufficient sleep on circadian rhythmicity and expression amplitude of the human blood transcriptome. Proc Natl Acad Sci U S A 110:E1132–E1141

Narasimamurthy R, Hatori M, Nayak SK, Liu F, Panda S, Verma IM (2012) Circadian clock protein cryptochrome regulates the expression of proinflammatory cytokines. Proc Natl Acad Sci U S A 109:12662–12667

Perez-Aso M, Feig JL, Mediero A, Aránzazu M, Cronstein BN (2013) Adenosine A2A receptor and TNF-α regulate the circadian machinery of the human monocytic THP-1 cells. Inflammation 36:152–162

Perry MG, Kirwan JR, Jessop DS, Hunt LP (2009) Overnight variations in cortisol, interleukin 6, tumour necrosis factor alpha and other cytokines in people with rheumatoid arthritis. Ann Rheum Dis 68:63–68

Prendergast BJ, Cable EJ, Patel PN, Pyter LM, Onishi KG, Stevenson TJ, Ruby NF, Bradley SP (2013) Impaired leukocyte trafficking and skin inflammatory responses in hamsters lacking a functional circadian system. Brain Behav Immun 32:94–104

Qin B, Deng Y (2015) Overexpression of circadian clock protein cryptochrome (CRY) 1 alleviates sleep deprivation-induced vascular inflammation in a mouse model. Immunol Lett 163:76–83

Rahman SA, Castanon-Cervantes O, Scheer FAJL, Shea SA, Czeisler CA, Davidson AJ, Lockley SW (2014) Endogenous circadian regulation of pro-inflammatory cytokines and chemokines in the presence of bacterial lipopolysaccharide in humans. Brain Behav Immun 47:4–13

Ramakrishnan SN, Muscat GEO (2006) The orphan Rev-erb nuclear receptors: a link between metabolism, circadian rhythm and inflammation? Nucl Recept Signal 4:e009

Sato S, Sakurai T, Ogasawara J, Takahashi M, Izawa T, Imaizumi K, Taniguchi N, Ohno H, Kizaki T (2014) A circadian clock gene, Rev-erbα, modulates the inflammatory function of macrophages through the negative regulation of Ccl2 expression. J Immunol 192:407–417

Schug TT, Li X (2011) Sirtuin 1 in lipid metabolism and obesity. Ann Med 43:198–211

Schuldenfrei A, Belton A, Kowalski J, Talbot CC, Di Cello F, Poh W, Tsai H-L, Shah SN, Huso TH, Huso DL et al (2011) HMGA1 drives stem cell, inflammatory pathway, and cell cycle progression genes during lymphoid tumorigenesis. BMC Genomics 12:549

Sherman H, Frumin I, Gutman R, Chapnik N, Lorentz A, Meylan J, le Coutre J, Froy O (2011) Long-term restricted feeding alters circadian expression and reduces the level of inflammatory and disease markers. J Cell Mol Med 15:2745–2759

Singh NN, Ramji DP (2008) Protein kinase CK2, an important regulator of the inflammatory response? J Mol Med 86:887–897

Straub RH, Buttgereit F, Cutolo M (2011) Alterations of the hypothalamic–pituitary–adrenal axis in systemic immune diseases – a role for misguided energy regulation. Clin Exp Rheumatol 29:S23–S31

Summa KC, Voigt RM, Forsyth CB, Shaikh M, Cavanaugh K, Tang Y, Vitaterna MH, Song S, Turek FW, Keshavarzian A (2013) Disruption of the circadian clock in mice increases intestinal permeability and promotes alcohol-induced hepatic pathology and inflammation. PLoS One 8:e67102

Thompson AAR, Walmsley SR, Whyte MKB (2014) A local circadian clock calls time on lung inflammation. Nat Med 20:809–811

Tong X, Buelow K, Guha A, Rausch R, Yin L (2012) USP2a protein deubiquitinates and stabilizes the circadian protein CRY1 in response to inflammatory signals. J Biol Chem 287:25280–25291

Tsoli M, Schweiger M, Vanniasinghe AS, Painter A, Zechner R, Clarke S, Robertson G (2014) Depletion of white adipose tissue in cancer cachexia syndrome is associated with inflammatory signaling and disrupted circadian regulation. PLoS One 9:e92966

Voigt RM, Forsyth CB, Keshavarzian A (2013) Circadian disruption: potential implications in inflammatory and metabolic diseases associated with alcohol. Alcohol Res 35:87–96

Wipfler P, Heikkinen A, Harrer A, Pilz G, Kunz A, Golaszewski SM, Reuss R, Oschmann P, Kraus J (2013) Circadian rhythmicity of inflammatory serum parameters: a neglected issue in the search of biomarkers in multiple sclerosis. J Neurol 260:221–227

Wirtz PH, Ehlert U, Emini L, Suter T (2008) Higher body mass index (BMI) is associated with reduced glucocorticoid inhibition of inflammatory cytokine production following acute psychosocial stress in men. Psychoneuroendocrinology 33:1102–1110

Wright KP, Drake AL, Frey DJ, Fleshner M, Desouza CA, Gronfier C, Czeisler CA (2015) Influence of sleep deprivation and circadian misalignment on cortisol, inflammatory markers, and cytokine balance. Brain Behav Immun 47:24–34

Yamakuchi M (2012) MicroRNA regulation of SIRT1. Front Physiol 3:68

Yang G, Wright CJ, Hinson MD, Fernando AP, Sengupta S, Biswas C, La P, Dennery PA (2014) Oxidative stress and inflammation modulate Rev-erbα signaling in the neonatal lung and affect circadian rhythmicity. Antioxid Redox Signal 21:17–32

Yoshida K, Hashiramoto A, Okano T, Yamane T, Shibanuma N, Shiozawa S (2013) TNF-α modulates expression of the circadian clock gene Per2 in rheumatoid synovial cells. Scand J Rheumatol 42:276–280

Yoshida K, Hashimoto T, Sakai Y, Hashiramoto A (2014) Involvement of the circadian rhythm and inflammatory cytokines in the pathogenesis of rheumatoid arthritis. J Immunol Res 2014:282495

Chapter 4
Circadian Rhythms and Cellular Networks in Depression and Associated Disorders

4.1 The Systemic Roles of Circadian Rhythms in Depression and Associated Disorders

As a common disorder in primary care, depressive disorders affect about 5 % people in developed countries every year (Friedman et al. 2011). Increasing evidences have indicated that circadian disruptions may be an essential factor in the pathophysiology of anxiety, major depressive disorder (MDD), bipolar disorders, and seasonal affective disorder (SAD) (Logan et al. 2015; Kronfeld-Schor and Einat 2012; Gorwood 2010; Pandi-Perumal et al. 2009). The severity of depressive symptoms has been closely related to the misalignment between the timing of sleep and the circadian pacemakers in patients with unipolar disorder (Courtet and Olié 2012).

The common symptoms in depression include difficulties in falling asleep, early morning awakenings, shorter sleep time and lower sleep efficiency, and changes in the rapid eye movement sleep. Other observations include the diurnal mood changes and alterations in the timing of temperature nadir and cortisol levels, which all indicate the associations between sleep disorders and depression (Kronfeld-Schor and Einat 2012; Boyce and Barriball 2010; Pandi-Perumal et al. 2009). Because it is a complex and multifactorial disease with multiple pathways involved, the simple modulation of serotonergic and noradrenergic neurotransmission has been found insufficient (Gorwood 2010). Studies at various levels on the basis of systems biology may help improve the understanding of such complexity.

Figure 4.1 provides a schematic illustration of the relevant pathways across various levels, from the environmental and life style factors to the change of mood including depression. These factors can affect the central clock in the suprachiasmatic nucleus (SCN) of the hypothalamus, rhythmic behaviors, metabolic processes, hypothalamus–pituitary–adrenal (HPA) axis, as well as immune responses. At the molecular and cellular levels, the networks of circadian genes may interact with neuro-endocrine-immune pathways, providing feedback to the higher levels (see Fig. 4.1). Circadian oscillations are produced and regulated via transcriptional,

© Springer International Publishing Switzerland 2015
Q. Yan, *Cellular Rhythms and Networks: Implications for Systems Medicine*,
SpringerBriefs in Cell Biology, DOI 10.1007/978-3-319-22819-8_4

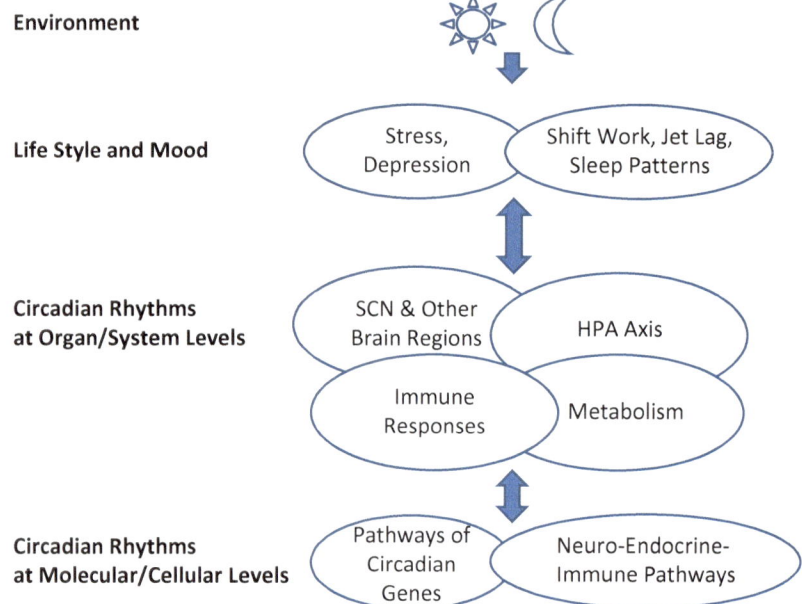

Fig. 4.1 A schematic illustration of the circadian–mood interactions across various levels

translational, and posttranslational feedback loops. Molecular circadian clocks in the cells control the timing of the expressions in a wide range of tissues in association with psychophysiological activities (Turek 2007). Such rhythms are coordinated all over the body through the functions of SCN (see Fig. 4.1).

4.1.1   Molecular Factors and Cellular Mechanisms of the Circadian–Mood Interactions

Depression and circadian disruptions may share a common etiology with lower cellular resilience and resistance to stressful events (Jakovljević 2011). Chronic stress may lead to the loss of rhythmic synchronization in the brain and provide the direct linkage between circadian disruptions and depressive behaviors. Circadian dysfunctions may have serious consequences with altered physiological functions in the brain and other tissue cells (Turek 2007). For example, lower amplitudes of molecular rhythms were observed in the extra-SCN brain areas among MDD patients (Logan et al. 2015). Unpredictable chronic mild stress (UCMS) may reduce the circadian amplitudes of activities and body temperature (Logan et al. 2015).

Older adults with a history of depression may have altered expressions of circadian genes including CLOCK, PER1, and BMAL1 (Gouin et al. 2010). Studies using mouse models showed that a selective decrease of CRY2 expression in the hippocampal tissue may be an important link between depression and circadian disruptions (Griesauer et al. 2014). Table 4.1 lists some examples of circadian-associated

Table 4.1 Examples of genes and pathways involved in the circadian–mood interactions

Genes	Changes/interactions	Associated mood disorders	References
BDNF	Lower expression in the hippocampus	Depressive-like responses	Fonken and Nelson (2013)
	Daily fluctuations, light therapy	Personality traits	Tirassa et al. (2012)
	Variants and 5-HTTLPR	Higher morning cortisol, depression in adolescents	Goodyer et al. (2010)
	Diurnal serum oscillations	Favorable treatment responses in MDD	Giese et al. (2014)
CRY2	Lower level in the hippocampal tissue	Anxiety, co-morbid depression	Griesauer et al. (2014)
	Depletion	Lower cognitive function, anxiety-related behaviors	De Bundel et al. (2013)
	SNPs rs10838524, rs7121611, rs7945565, rs1401419	Dysthymia, winter depression, bipolar type 1	Kovanen et al. (2013)
	Lower levels	Bipolar disorder	Sjöholm et al. (2010)
	Variant	Depression	Lavebratt et al. (2010a)
PER2	Higher levels in the amygdala	Antipsychotic drug quetiapine (QTP)	Moriya et al. (2014)
	Genetic variations	Depression vulnerability	Lavebratt et al. (2010b)
	Mutations in CLOCK or PER2	Mania-like behaviors	Kennaway (2010)
	Variants in CLOCK, NPAS2, PER2, PER3	Altered sleep, inertia, alcohol consumption	Gamble et al. (2011)
	Lower expression in the SCN	Depressive-like behavior	Jiang et al. (2011)
RORA	SNPs rs12912233, rs880626, SCN1A SNP rs3812718	Autism, depression, epilepsy	Haerian et al. (2015)
	SNP rs2028122	Depressive disorder	Partonen (2012)
	SNP rs12912233	The trait depression	Terracciano et al. (2010)
	TIMELESS, ARNTL (also known as BMAL1), NR1D1	Depression with early morning awakening	Utge et al. (2010)
TIMELESS	SNPs rs7486220, rs1082214; ARNTL, RORA, NR1D1	Depression and sleep disturbances	Utge et al. (2010)
	Variants	Mood disorders	Mendlewicz (2009)
	CLOCK, PER1, PER3	Chronic schizophrenia, sleep disturbances	Lamont et al. (2007)
	BMAL1, PER3	Bipolar disorder	Kato (2007)

molecular factors and cellular networks that may have important roles in depression and chronotherapy. Some of these factors are discussed in details below. A more complete list can be found in the Database of Biological Rhythms (DBBR 2015).

4.1.1.1   PER2 and Associated Networks

The PER2 protein has been considered an integrator and mediator on both of the input and output sides of the circadian clock (Ripperger and Albrecht 2012). The PER proteins interact with many other proteins to transfer the oscillator timing information. They are involved in the synchronization of many rhythmic processes. Chronic unpredictable stress (CUS) has been found to induce lower levels of PER2 expression in the SCN (Jiang et al. 2011). This change may be associated with the development of depressive-like behaviors.

Genetic variants in PER2, PER3, CLOCK, and NPAS2 have been related to various behavioral changes including alcohol/caffeine consumptions, sleepiness, altered sleep phases, as well as inertia (Gamble et al. 2011). The gene–environment interactions may contribute to these alterations as often seen in the cases of shift work (Gamble et al. 2011; also see Fig. 4.1). In a study among a Swedish population, a PER2 genetic variation was closely related to the vulnerability to depression (Lavebratt et al. 2010b). Such genetic risk seemed to be independent from negative life events that are often associated with the risks for depression. In addition, studies using mouse models found that mice having mutations in the CLOCK or PER2 genes spent less time immobile in swim tests, indicating behaviors mimicking mania (Kennaway 2010). In humans, changes in such clock genes may have important roles in depression.

These mechanisms indicate that novel antidepressant drugs can be designed to target the altered structures and functions in the circadian timing systems (Kennaway 2010). Such correlations may also help explain the mechanisms of the antidepressant effects of some drugs. For example, the antipsychotic drug quetiapine (QTP) has been accepted as a potential therapeutic agent for depression with unclear mechanisms. A study using a mouse model found that QTP could promote the PER2 expression in the mouse amygdala (Moriya et al. 2014). Such findings indicate the potential effects of QTP on the circadian systems with impacts on the mood symptoms.

4.1.1.2   RORA and Associated Networks

The genes RAR-related orphan receptors A (RORA), B (RORB), and voltage-gated sodium channel type 1 (SCN1A) may have the key roles in the regulation of the circadian clocks. Studies have found the linkages between circadian genes and gender-dependent depression with sleep disturbance. For example, together with the clock genes TIMELESS, ARNTL (also known as BMAL1), and NR1D1, altered RORA has been related to depression with early morning awakening among male subjects (Utge et al. 2010).

In addition, genetic variants in RORA and RORB have been associated with the susceptibility to autism and depression (Haerian et al. 2015). A meta-analysis about genotypes and psychological phenotypes found a strong correlation between the RORA SNP rs12912233 and trait depression (Terracciano et al. 2010). Such analyses support the correlation between certain genotypes and various phenotypes from common personality traits to psychiatric disorders.

In another example, the associations between depression and the genetic variants of RORA (rs2028122) and CRY1 (rs2287161) were established (Partonen 2012). Furthermore, a study among Malaysian Chinese samples found that the RORA SNPs rs12912233, rs880626, and the SCN1A SNP rs3812718 may have a synergistic effect in the risks for epilepsy (Haerian et al. 2015).

Certain individuals have higher risks for the development of depression including the feelings of guilt, sadness, and hopelessness. The genetic variants such as those in the RORA associated pathways can be identified to recognize such individual differences in psychiatric disorders for the practice of personalized medicine.

4.1.2 Systemic Factors in the Circadian–Mood Interactions

On the basis of systems biology, factors at various levels can be identified in the correlation between circadian disruptions and depression. Specifically, depression has been closely related to the alterations in the HPA axis (Gudmand-Hoeyer et al. 2014). A study using the chronic mild stress (CMS) rat model of depression identified abnormal diurnal rhythms and corticosterone levels among stress-susceptible animals in comparison with stress-resilient animals (Christiansen et al. 2012). Such disturbance of the HPA axis rhythmicity was observed during the beginning stages of depression.

As suggested by a mechanistic nonlinear model, both circadian and ultradian oscillations of the hormone concentrations (such as ACTH and cortisol) are associated with the activities of the HPA-axis (Gudmand-Hoeyer et al. 2014). While the ultradian rhythm originates from the hippocampus region, the 24-h oscillations are regulated by the central and peripheral clocks. The analyses of such mechanisms at various levels may be helpful for the identification of more precise biomarkers and therapeutic targets (Gudmand-Hoeyer et al. 2014).

Moreover, factors such as aging, shift work, and human–environment interactions are also important. For instance, altered lighting conditions and lifestyles may lead to the disturbance in circadian rhythms with higher risks for mood disorders such as impulsivity, mania, and depression (Salgado-Delgado et al. 2011). A study of 1714 middle-aged and elderly participants showed that the fragmentation of the 24-h activity rhythm is closely related to aging, depression and anxiety (Luik et al. 2015). Disorganization in the circadian systems caused by these factors may contribute to neurobiological dysfunctions (Turek 2007). Together with altered endocrine and metabolic functions, all of these factors need to be considered in studying depression and associated disorders (see Fig. 4.1).

4.2 Depression and Systemic Circadian Profiles: Implications for Personalized Medicine

4.2.1 Prediction, Prevention, and Personalized Interventions for Depression

The analysis of systemic circadian profiles in mood changes would be helpful for the prediction of not only the risks but also therapeutic responses. Because circadian disturbances are significant in many patients with major depression, the identification of such subgroups may be meaningful for personalized diagnosis and treatment (Hickie et al. 2013). Objective measurements can be used to establish the systemic circadian profiles including various components such as daytime activities, nighttime sleep patterns, the timing of circadian-dependent physiology, as well as the sleep–wake cycles.

Specifically, the sleep–wake cycle is an essential component of the circadian profiles. People with disruptions in circadian patterns and sleep–wake cycles often have mental problems (Jakovljević 2011). Insomnia and circadian disturbances are remarkable properties of MDD, referring to the critical role of the altered melatonin production (Pandi-Perumal et al. 2009). These observations indicate that disturbed circadian and sleep–wake cycles may be the key risk factors for the development, exacerbation, recurrence, and unfavorable outcomes of depression and other mental problems (Jakovljević 2011; Turek 2007).

The pattern analysis of the circadian profiles can be very useful for the prediction and prevention of various psychiatric disorders. For example, the circadian profiles of patients with MDD and panic disorder (PD) have been found different from those of healthy people even several years before the clinical onset of the disorders (Bersani et al. 2012). Such behavioral changes often occur during the premorbid age between 12 and 20 years old, including alterations in the patterns of falling asleep, awakening, and cognitive functioning. The early detection of such changes in the circadian profiles may be helpful for taking preventive measures among those with high psychiatric vulnerabilities (Bersani et al. 2012).

4.2.2 The Circadian–Mood Interactions in Cancer and Other Complex Diseases

The circadian profiles may also provide the linkages between mood disorders and other complex diseases. For instance, cancer patients often have abnormal circadian rhythms associated with low quality of life. A study of 68 breast cancer patients in Japan confirmed that compared to the cancer-free controls, the patients had more disrupted circadian patterns, worse sleep quality, more symptoms of depression, and more fatigue (Ancoli-Israel et al. 2014).

In another study of 219 US patients with breast cancer (stages I–IIIA), circadian disruptions were found as a common feature (Berger et al. 2010). In addition, such alterations were related to stress, fatigue, and symptoms of depression not only during chemotherapy but also in the recovery stages. On the other hand, effective interventions led to more robust rhythms with reductions in fatigue and depression-associated symptoms, as well as better body mass index and performance (Berger et al. 2010).

In addition to cancers, circadian profiles and depression have been associated with other complex disorders such as coronary heart disease. In a study of 88 patients with suspected coronary artery disease (CAD), the association was evaluated between depression and cortisol because cortisol has a critical role in the development of CAD (Bhattacharyya et al. 2008). The study found that the cortisol slope over the day was flatter among CAD patients who were more depressed. These patients had reduced levels of cortisol early in the day with increased levels of cortisol in the evening (Bhattacharyya et al. 2008). The study indicates that the flatter cortisol rhythms may result in the progression of coronary atherosclerosis among the more depressed CAD patients.

Furthermore, altered circadian profiles have been related to aging-associated disorders. A study of 48 community-dwelling older adults observed significant lower levels of salivary cortisol, awakening response, and a flatter diurnal rhythm (Heaney et al. 2010). Such changes were also correlated with anxiety and depression. In the neurodegenerative disorder Huntington's disease, the properties of night-time sleep impairment and delayed sleep phases have been linked to circadian alterations, lower cognitive and functional performance, as well as depression (Aziz et al. 2010). Because circadian disruptions have been deemed as one of the risk factors for depression and associated disorders, the adjustment of the altered rhythms may provide novel avenues for more effective therapeutic strategies.

4.2.3   Systemic Circadian Profiles and Chronotherapy for Depression

Currently, available antidepressant therapies such as the selective serotonin reuptake inhibitors are unsatisfactory with remarkable limitations. They are effective in just about half of the patients, with slow onset of action while the delayed responses may take several weeks to achieve (Friedman et al. 2011; Lader 2007). Other problems include the low rates of tolerability, withdrawal symptoms, compliance problems, and high rates of recurrence (Gorwood 2010; Lader 2007).

Systemic circadian profiles can be established on the basis of the parameters at various levels, including molecular, cellular, and lifestyle factors (see Sects. 4.1 and 4.2). Such method would enable the discovery of more accurate biomarkers for diagnosis and prognosis, early prediction, and personalized treatment. For example, the circadian rhythms of heart rate, locomotor activity, and temperature have been suggested as potential parameters for the objective diagnosis of depression and early prediction of therapeutic responses (Friedman et al. 2011).

Conventional studies of the treatment of depression have been focusing on the neurotransmitter disturbances. However, recent discoveries about chronotherapeutics and the resynchronizing properties of antidepressants on the circadian rhythms may offer innovative approaches for more integrative therapies. Antidepressants may have effects on sleep deprivation, insomniac depressives, and circadian patterns (Kronfeld-Schor and Einat 2012; Lader 2007). Chronotherapy such as scheduled medication may also affect the therapeutic efficiency, showing that the circadian systems may be a crucial target for the better treatment of depression (Quera Salva et al. 2011; Lader 2007).

For instance, chronobiotic agents and bright light therapy have been suggested helpful for the treatment of seasonal affective disorders and mood disturbances via the regulation of the circadian systems (Quera Salva et al. 2011; Boyce and Barriball 2010). Interpersonal and social rhythm therapy (IPSRT) focuses on the management of interpersonal relationships as well as the regulation of social cues including the adjustment of the timing of sleep/wake and meals (Boyce and Barriball 2010). IPSRT has been considered both a preventive and an acute intervention for bipolar depression (Frank 2007).

Moreover, potential pharmaceutical medications may be developed as antidepressants via the regulation of the circadian systems. For example, the novel melatonergic compounds including the melatonin agonist agomelatine have effects on sleep quality and alertness at awakening (Quera Salva et al. 2011). They may benefit those with MDD, SAD, or bipolar disorder (Quera Salva et al. 2011; Gorwood 2010; Pandi-Perumal et al. 2009). Experiments in animal models and placebo-controlled trials about agomelatine have shown quick responses and rapid onset of action (Gorwood 2010; Pandi-Perumal et al. 2009). Other benefits include high tolerability, low rates of relapse, as well as remarkable improvements of the major symptoms of anxiety and depression. It has also been found to help reconstruct the amplitude of the circadian rhythms and the sleep–wake cycle (Gorwood 2010; Pandi-Perumal et al. 2009). Such compounds with chronobiotic features of resynchronization may provide more effective therapeutic strategies for depression and other mental problems with similar etiology.

References

Ancoli-Israel S, Liu L, Rissling M, Natarajan L, Neikrug AB, Palmer BW, Mills PJ, Parker BA, Sadler GR, Maglione J (2014) Sleep, fatigue, depression, and circadian activity rhythms in women with breast cancer before and after treatment: a 1-year longitudinal study. Support Care Cancer 22:2535–2545

Aziz NA, Anguelova GV, Marinus J, Lammers GJ, Roos RAC (2010) Sleep and circadian rhythm alterations correlate with depression and cognitive impairment in Huntington's disease. Parkinsonism Relat Disord 16:345–350

Berger AM, Wielgus K, Hertzog M, Fischer P, Farr L (2010) Patterns of circadian activity rhythms and their relationships with fatigue and anxiety/depression in women treated with breast cancer adjuvant chemotherapy. Support Care Cancer 18:105–114

Bersani G, Bersani FS, Prinzivalli E, Limpido L, Marconi D, Valeriani G, Colletti C, Anastasia A, Pacitti F (2012) Premorbid circadian profile of patients with major depression and panic disorder. Riv Psichiatr 47:407–412

Bhattacharyya MR, Molloy GJ, Steptoe A (2008) Depression is associated with flatter cortisol rhythms in patients with coronary artery disease. J Psychosom Res 65:107–113

Boyce P, Barriball E (2010) Circadian rhythms and depression. Aust Fam Physician 39:307–310

Christiansen S, Bouzinova EV, Palme R, Wiborg O (2012) Circadian activity of the hypothalamic–pituitary–adrenal axis is differentially affected in the rat chronic mild stress model of depression. Stress 15:647–657

Courtet P, Olié E (2012) Circadian dimension and severity of depression. Eur Neuropsychopharmacol 22(Suppl 3):S476–S481

DBBR (2015) The database of biological rhythms. http://pharmtao.com/health/biological-rhythms-database/. Accessed 1 June 2015

De Bundel D, Gangarossa G, Biever A, Bonnefont X, Valjent E (2013) Cognitive dysfunction, elevated anxiety, and reduced cocaine response in circadian clock-deficient cryptochrome knockout mice. Front Behav Neurosci 7:152

Fonken LK, Nelson RJ (2013) Dim light at night increases depressive-like responses in male C3H/HeNHsd mice. Behav Brain Res 243:74–78

Frank E (2007) Interpersonal and social rhythm therapy: a means of improving depression and preventing relapse in bipolar disorder. J Clin Psychol 63:463–473

Friedman A, Shaldubina A, Flaumenhaft Y, Weizman A, Yadid G (2011) Monitoring of circadian rhythms of heart rate, locomotor activity, and temperature for diagnosis and evaluation of response to treatment in an animal model of depression. J Mol Neurosci 43:303–308

Gamble KL, Motsinger-Reif AA, Hida A, Borsetti HM, Servick SV, Ciarleglio CM, Robbins S, Hicks J, Carver K, Hamilton N et al (2011) Shift work in nurses: contribution of phenotypes and genotypes to adaptation. PLoS One 6, e18395

Giese M, Beck J, Brand S, Muheim F, Hemmeter U, Hatzinger M, Holsboer-Trachsler E, Eckert A (2014) Fast BDNF serum level increase and diurnal BDNF oscillations are associated with therapeutic response after partial sleep deprivation. J Psychiatr Res 59:1–7

Goodyer IM, Croudace T, Dudbridge F, Ban M, Herbert J (2010) Polymorphisms in BDNF (Val66Met) and 5-HTTLPR, morning cortisol and subsequent depression in at-risk adolescents. Br J Psychiatry 197:365–371

Gorwood P (2010) Restoring circadian rhythms: a new way to successfully manage depression. J Psychopharmacol (Oxford) 24:15–19

Gouin J-P, Connors J, Kiecolt-Glaser JK, Glaser R, Malarkey WB, Atkinson C, Beversdorf D, Quan N (2010) Altered expression of circadian rhythm genes among individuals with a history of depression. J Affect Disord 126:161–166

Griesauer I, Diao W, Ronovsky M, Elbau I, Sartori S, Singewald N, Pollak DD (2014) Circadian abnormalities in a mouse model of high trait anxiety and depression. Ann Med 46:148–154

Gudmand-Hoeyer J, Timmermann S, Ottesen JT (2014) Patient-specific modeling of the neuroendocrine HPA-axis and its relation to depression: ultradian and circadian oscillations. Math Biosci 257:23–32

Haerian BS, Sha'ari HM, Tan HJ, Fong CY, Wong SW, Ong LC, Raymond AA, Tan CT, Mohamed Z (2015) RORA gene rs12912233 and rs880626 polymorphisms and their interaction with SCN1A rs3812718 in the risk of epilepsy: a case–control study in Malaysia. Genomics 105:229–236

Heaney JLJ, Phillips AC, Carroll D (2010) Ageing, depression, anxiety, social support and the diurnal rhythm and awakening response of salivary cortisol. Int J Psychophysiol 78:201–208

Hickie IB, Naismith SL, Robillard R, Scott EM, Hermens DF (2013) Manipulating the sleep–wake cycle and circadian rhythms to improve clinical management of major depression. BMC Med 11:79

Jakovljević M (2011) Agomelatine as chronopsychopharmaceutics restoring circadian rhythms and enhancing resilience to stress: a wishfull thinking or an innovative strategy for superior management of depression? Psychiatr Danub 23:2–9

Jiang W-G, Li S-X, Zhou S-J, Sun Y, Shi J, Lu L (2011) Chronic unpredictable stress induces a reversible change of PER2 rhythm in the suprachiasmatic nucleus. Brain Res 1399:25–32

Kato T (2007) Molecular genetics of bipolar disorder and depression. Psychiatry Clin Neurosci 61:3–19

Kennaway DJ (2010) Clock genes at the heart of depression. J Psychopharmacol (Oxford) 24:5–14

Kovanen L, Kaunisto M, Donner K, Saarikoski ST, Partonen T (2013) CRY2 genetic variants associate with dysthymia. PLoS One 8, e71450

Kronfeld-Schor N, Einat H (2012) Circadian rhythms and depression: human psychopathology and animal models. Neuropharmacology 62:101–114

Lader M (2007) Limitations of current medical treatments for depression: disturbed circadian rhythms as a possible therapeutic target. Eur Neuropsychopharmacol 17:743–755

Lamont EW, Legault-Coutu D, Cermakian N, Boivin DB (2007) The role of circadian clock genes in mental disorders. Dialogues Clin Neurosci 9:333–342

Lavebratt C, Sjöholm LK, Soronen P, Paunio T, Vawter MP, Bunney WE, Adolfsson R, Forsell Y, Wu JC, Kelsoe JR et al (2010a) CRY2 is associated with depression. PLoS One 5, e9407

Lavebratt C, Sjöholm LK, Partonen T, Schalling M, Forsell Y (2010b) PER2 variation is associated with depression vulnerability. Am J Med Genet B Neuropsychiatr Genet 153B:570–581

Logan RW, Edgar N, Gillman AG, Hoffman D, Zhu X, McClung CA (2015) Chronic stress induces brain region-specific alterations of molecular rhythms that correlate with depression-like behavior in mice. Biol Psychiatry 78:249–258

Luik AI, Zuurbier LA, Direk N, Hofman A, Van Someren EJW, Tiemeier H (2015) 24-Hour activity rhythm and sleep disturbances in depression and anxiety: a population-based study of middle-aged and older persons. Depress Anxiety 32:684–692

Mendlewicz J (2009) Disruption of the circadian timing systems: molecular mechanisms in mood disorders. CNS Drugs 23(Suppl 2):15–26

Moriya S, Tahara Y, Sasaki H, Hamaguchi Y, Kuriki D, Ishikawa R, Ishigooka J, Shibata S (2014) Effect of quetiapine on Per1, Per2, and Bmal1 clock gene expression in the mouse amygdala and hippocampus. J Pharmacol Sci 125:329–332

Pandi-Perumal SR, Moscovitch A, Srinivasan V, Spence DW, Cardinali DP, Brown GM (2009) Bidirectional communication between sleep and circadian rhythms and its implications for depression: lessons from agomelatine. Prog Neurobiol 88:264–271

Partonen T (2012) Clock gene variants in mood and anxiety disorders. J Neural Transm 119: 1133–1145

Quera Salva MA, Hartley S, Barbot F, Alvarez JC, Lofaso F, Guilleminault C (2011) Circadian rhythms, melatonin and depression. Curr Pharm Des 17:1459–1470

Ripperger JA, Albrecht U (2012) The circadian clock component PERIOD2: from molecular to cerebral functions. Prog Brain Res 199:233–245

Salgado-Delgado R, Tapia Osorio A, Saderi N, Escobar C (2011) Disruption of circadian rhythms: a crucial factor in the etiology of depression. Depress Res Treat 2011:839743

Sjöholm LK, Backlund L, Cheteh EH, Ek IR, Frisén L, Schalling M, Osby U, Lavebratt C, Nikamo P (2010) CRY2 is associated with rapid cycling in bipolar disorder patients. PLoS One 5, e12632

Terracciano A, Tanaka T, Sutin AR, Sanna S, Deiana B, Lai S, Uda M, Schlessinger D, Abecasis GR, Ferrucci L et al (2010) Genome-wide association scan of trait depression. Biol Psychiatry 68:811–817

Tirassa P, Iannitelli A, Sornelli F, Cirulli F, Mazza M, Calza A, Alleva E, Branchi I, Aloe L, Bersani G et al (2012) Daily serum and salivary BDNF levels correlate with morning–evening personality type in women and are affected by light therapy. Riv Psichiatr 47:527–534

Turek FW (2007) From circadian rhythms to clock genes in depression. Int Clin Psychopharmacol 22(Suppl 2):S1–S8

Utge SJ, Soronen P, Loukola A, Kronholm E, Ollila HM, Pirkola S, Porkka-Heiskanen T, Partonen T, Paunio T (2010) Systematic analysis of circadian genes in a population-based sample reveals association of TIMELESS with depression and sleep disturbance. PLoS One 5, e9259

Chapter 5
Circadian Rhythms and Cellular Networks in Cardiovascular Diseases

5.1 A Systemic Overview of the Circadian–Cardiovascular Interactions

The elucidation of cellular processes and rhythms are critical for understanding the mechanisms of cardiovascular diseases (CVDs). The operative configuration of the cardiovascular system displays evident circadian rhythmicity. Circadian rhythms have critical roles in both of the cardiovascular physiological and pathological processes. For example, diurnal oscillations are apparent in heart rates, blood pressure, and endothelial functions. Diurnal variations are also the properties of the onset of many CVDs including acute coronary syndrome, atrial arrhythmia, and subarachnoid hemorrhage (Takeda and Maemura 2011). The disruption of the circadian patterns, the loss of the synchronization, and the dysfunctions in relevant pathways may disturb the adaptive tissue responses to stimuli. Such changes may lead to cardiovascular disorders including hypertension, myocardial infarction, and atherosclerosis (see Fig. 5.1).

Figure 5.1 provides a systemic overview of the interactions between the circadian system and the cardiovascular system, as well as the possible pathways toward CVDs at various levels. The environmental factors such as the light/dark cycles and psychosocial/physiological stress may influence the central clock in the suprachiasmatic nucleus (SCN) of the hypothalamus and other parts of the neuroendocrine systems. In addition, cardiovascular tissues have their own peripheral clocks that can directly control the circadian oscillations of cardiovascular functions. Together with metabolic cycles and factors, autonomic and/or hormonal factors may synchronize and regulate the peripheral oscillations in the cardiovascular tissues including the heart and arteries (see Fig. 5.1).

At the molecular and cellular levels, the circadian pathways including clock genes CLOCK/BMAL1, PER/CRY, RORA, and REV-ERBα (also known as NR1D1) may interact with other physiological pathways (see Fig. 5.1). The associated networks include the lipid metabolic pathways involving PPARs and the

© Springer International Publishing Switzerland 2015   49
Q. Yan, *Cellular Rhythms and Networks: Implications for Systems Medicine*,
SpringerBriefs in Cell Biology, DOI 10.1007/978-3-319-22819-8_5

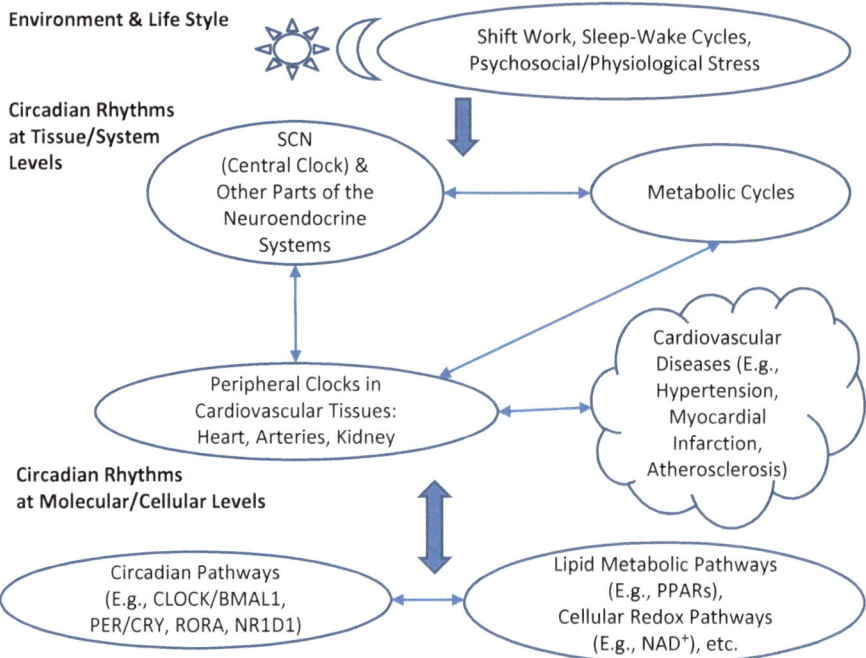

Fig. 5.1 A systemic overview of the circadian–cardiovascular interactions at various levels

cellular redox conditions involving NAD⁺ (Kohsaka et al. 2012). A more detailed discussion on some examples of the genes and pathways involved in the circadian–cardiovascular interactions will be provided in the following sections.

5.1.1 The Circadian–Cardiovascular Interactions at the Molecular and Cellular Levels

The cardiovascular system responds to environmental stimuli with circadian patterns. Such patterns are mediated via complex interactions between the extracellular factors such as neuro-humoral elements and intracellular factors such as the clock genes (Dominguez-Rodriguez et al. 2010). To maintain the normal cardiac organ functions, the synchrony between the endogenous and outside clocks and the orchestration among molecular rhythms in the cells are necessary. Diurnal variations have been observed in gene expressions in the heart and blood cells. These molecular and cellular factors can become potential targets with implications for more effective chronotherapy.

For example, one factor with a critical role in the cardiovascular functions is melatonin. It is a neuro-hormone that exhibits a diurnal oscillation with chronobiotic and epigenetic regulatory functions. As an antioxidant with anti-inflammatory effects, melatonin is essential in the synchronization of molecular circadian rhythms in both SCN and peripheral tissues (Dominguez-Rodriguez et al. 2010). The blood melatonin rhythm is crucial in the maintenance of various cardiovascular functions such as the daily oscillations in blood pressure.

Another important modulator of the circadian rhythmicity in the heart is the neuropeptide vasoactive intestinal peptide (VIP). VIP is involved in many circadian-associated activities, including the SCN electrical processes and wheel running behaviors in mice (Schroeder et al. 2011). SCN neurons expressing VIP may be involved in the circadian control of cardiac operations via the interactions with the autonomic centers. In VIP-deficient mice in the light–dark conditions, weak rhythms in the heart rate, body temperature, and cage activity were observed (Schroeder et al. 2011). Such mice also lost the regular circadian rhythms in the heart rate in the condition of constant darkness.

Other molecules have significant impacts too. For instance, the PER2 gene is one of the core elements of the molecular circadian clocks. Studies using mouse models found that mutations of the PER2 gene may lead to an altered dipping of blood pressure (BP) and heart rate under the conditions of both light–dark cycles and constant darkness (Vukolic et al. 2010). The changes also led to shorter circadian periods under the condition of constant darkness. Such observations demonstrate the important roles of the molecular connections between the circadian modulation and cardiovascular pathogenesis.

In addition, the expression cycles of metabolic genes in cardiovascular tissues may interact with the transcriptional regulation of the circadian genes in a bidirectional manner. While the metabolic cycles have impacts on the circadian rhythms of cardiovascular functions, the circadian systems may integrate both metabolic and cardiovascular functions (Kohsaka et al. 2012; also see Fig. 5.1).

Furthermore, networks such as the oxidative stress pathways may also be critical. For example, disrupted cardiovascular circadian patterns have been widely observed in chronic heart failure (CHF). A study using a post myocardial infarction (MI) CHF mouse model showed circadian disruptions in arterial pressure, heart rate, and baroreflex sensitivity after MI (Mousa et al. 2014). Such changes occurred together with the upregulation of the angiotensin II type 1 receptors (AT1R) and gp91(phox) proteins in the brainstem. These changes indicate that central oxidative stress may be involved in the circadian cardiovascular alterations (Mousa et al. 2014). In the next section, more examples of the molecular and cellular factors will be provided.

5.1.2   Examples of Circadian-Associated Molecular Factors and Cellular Networks

Table 5.1 lists some examples of circadian-associated molecular factors and cellular networks that may have important roles in CVDs and chronotherapy. Some of these factors are discussed in details below. A more complete list can be found in the Database of Biological Rhythms (DBBR 2015).

Table 5.1 Examples of circadian-associated genes and pathways involved in cardiovascular diseases

Genes	Changes/interactions	Roles in cardiovascular diseases	References
ACE	Cardiovascular rhythms, PER2	Locomotor activity, hypertension	Herichová et al. (2013)
	DD genotype	Impaired BP variations in type 2 diabetes	Czupryniak et al. (2008)
	D allele	Left ventricular mass index in hypertension	Kulah et al. (2007)
	D allele	Hypertension	Spiering et al. (2005)
	ACE inhibitors	Hypertension, chronotherapy	Hermida et al. (2011)
	ACE inhibitor dosing time	Chronotherapy, cardiac hypertrophy in hypertension	Ohmori and Fujimura (2005)
ADIPOQ	Low adiponectin levels	Liver steatosis, insulin resistance, hypertension	Fallo et al. (2008)
	Hypoxic stress, nocturnal reduction in adiponectin levels	Obstructive sleep apnea-hypopnea syndrome, atherosclerotic diseases	Nakagawa et al. (2008)
NR1D1	Deletion	Atherosclerotic lesions	Ma et al. (2013)
	Ligands targeting the circadian rhythms	Sleep disorders, obesity, dyslipidemia, hyperglycemia	Solt et al. (2012)
	Interactions with other nuclear receptors	Abnormal lipid metabolism, vascular inflammation	Duez and Staels (2008)
	Interactions with other nuclear receptors	Atherosclerosis, cardiovascular risks	Wang et al. (2006)
PPARG	Cardiovascular rhythms, BMAL1, Wnt/beta-catenin pathways	Arrhythmogenic right ventricular cardiomyopathy, type 2 diabetes	Lecarpentier et al. (2014)
	Deletion	Suppressed oscillations in food intake, cardiovascular functions	Yang et al. (2012)
	Plasminogen activator inhibitor-1 (PAI-1)	Ketogenic diet (KD)-induced hypofibrinolysis	Oishi et al. (2010)
	Dysfunctions of clock genes	Hyperphagia, obesity, myocardial infarction, sudden cardiac death	Lecarpentier et al. (2010)
	Cardiovascular rhythms, BMAL1	Altered oscillations in BP and heart rate	Wang et al. (2008)

5.1.2.1   ACE and Chronotherapy

Circadian rhythms are critical for cardiovascular physiology and the timing of adverse cardiovascular events (Martino et al. 2007). Because the angiotensin-converting enzyme (ACE) is one of the core genes in the hypertrophic pathways, ACE inhibitors may have key roles in the circadian regulation of BP and chronotherapy (Martino et al. 2007; Hermida et al. 2011). The insertion/deletion (ACE I/D) polymorphisms may be closely associated with the 24-h BP profiles in hypertensive patients (Spiering et al. 2005). Genetic variants in ACE may have key roles in the impaired circadian BP variation, endothelial dysfunction, and higher mortality rate among normotensive type 2 diabetes patients (Czupryniak et al. 2008). In a study among the Turkish hypertensive population, the ACE polymorphisms were found to affect the 24-h ambulatory BP measurement values and the risk for left ventricular mass (Kulah et al. 2007).

On the basis of such observations, frequent measurements of the ambulatory BP monitoring (ABPM) have been suggested. The 24-h BP patterns have been connected to the progressive damages of target tissues and higher risks of cardiac and cerebrovascular events (Hermida et al. 2011). Specifically, blunted asleep BP decline may be associated with higher rates of fatal and nonfatal cardiovascular disorders. These mechanisms indicate that better descriptions of the 24-h BP patterns and better BP control during nighttime sleep may be helpful for the prevention of cardiovascular events.

Because the dosing of ACE inhibitors in the inactive span may be more effective with lower adverse reactions, chronotherapy has been suggested as a cost-effective method for hypertension and other cardiovascular disorders (Hermida et al. 2011). The applications of ACE inhibitors in the treatments of hypertension and heart failure may have adverse effects including angioedema and dry cough. However, changing the dosing time from morning to evening may lower the severity and frequency of the drug-caused dry cough (Ohmori and Fujimura 2005). Moreover, administered at different time-of-day in the morning or in the evening, different effects have been observed for the ACE inhibitors including benazepril, captopril, enalapril, perindopril, and quinapril (Hermida et al. 2011).

5.1.2.2   NR1D1 and Associated Networks

The synchronization of the rhythms in behaviors and metabolic activities is critical for the maintenance of cardiovascular health. NR1D1 has an integral role in the regulation of the operating rhythms in both behaviors and metabolism. It is a key gene in circadian timing, lipid and lipoprotein metabolism, adipogenesis, as well as vascular inflammation (Ma et al. 2013; Duez and Staels 2008).

As a regulator of cardiovascular risk factors, NR1D1 interacts with other nuclear receptors in the energy homeostasis pathways (Duez and Staels 2008; Wang et al. 2006). Such interactions have an essential role in the synchronization of metabolic processes and circadian activities. For instance, NR1D1 may be a modulator of atherogenesis in hematopoietic cells as its deficiency has been associated with atherosclerotic lesions (Ma et al. 2013). In addition, the disturbance of sleep homeostasis may cause the circadian disruption and various disorders including CVDs, obesity, and cognitive destruction. Many circadian genes can be affected by insufficient sleep, including NR1D1, NR1D2, PER1, PER2, PER3, CRY2, CLOCK, and RORA (Möller-Levet et al. 2013).

The understanding of the roles of NR1D1 in metabolic and cardiovascular activities may contribute to the development of novel therapeutics. For example, a NR1D1 agonist was found to reduce fat mass and hyperglycemia in diet-induced obese mice (Solt et al. 2012). Synthetic NR1D1 ligands may have effects on the circadian expression patterns of the clock genes in the hypothalami, as well as the metabolic genes in the liver, skeletal muscles and adipose tissues (Solt et al. 2012). Leading to higher energy expenditure, synthetic NR1D1 ligands targeting the circadian patterns may be useful for the treatment of both metabolic and cardiovascular disorders.

5.1.3   The Cell Cycle Interactions and Pathways

Together with the circadian rhythms, cell cycle also has an essential role in cardiovascular functions. Various elements are involved in such processes, such as the cyclins, the cyclin-dependent kinases (CDKs) and the cyclin-dependent kinase inhibitors (CKIs) (Boehm and Nabel 2003). These cell cycle regulatory molecules are the downstream targets of the mitogen-activated kinase (MAPK) and endothelin-1 (ET-1) pathways. These pathways are critical in the pathogenesis of CVDs caused by various factors including hypoxia, infections, and cardiomyopathy (Petkova et al. 2000). The elucidation of the circuitries such as the cyclin-CDK-CKI interactions may be helpful for the development of more effective therapeutic agents for CVDs (Boehm and Nabel 2003).

In addition, the cell cycle marker p53 in epicardial stromal cells may have important roles in adipocyte enlargement among obese patients with CVDs (Agra et al. 2014). Cell cycle checkpoint signal transduction cascades have been found important for the early diagnosis of cellular senescence, cardiovascular and neurodegenerative diseases, as well as cancer (Golubnitschaja 2007). The reprogramming of the cell cycle machinery has been suggested as effective strategies for the treatment of CVDs including coronary artery disease (Bicknell and Brooks 2008).

5.1.4   The Central and Peripheral Clocks Associated
with the Cardiovascular Functions

Rhythmic cardiovascular functions such as the 24-h BP patterns are mainly operated by the circadian clock systems (Kohsaka et al. 2012). The light/dark cycle-entrained central pacemaker in the SCN can convey the orchestrating signals to the peripheral clocks in tissues including the heart and blood vessels (see Fig. 5.1). The environmental factors that have the strongest impacts on mammalian circadian clocks include light, sound, temperature, physical activities such as the sleep/wake transitions, shift work, as well as food intake (Young and Bray 2007).

The peripheral clocks exist in various cardiovascular tissues and cells with their own intrinsic biological rhythms, including the cardiomyocyte, aortic tissue, vascular smooth muscle cells, and vascular endothelial cells (Takeda and Maemura 2011). The peripheral rhythms are also influenced by the temporal alterations in energy homeostasis (Kohsaka et al. 2012).

The fact that the clock genes are regulated by circadian oscillations not just in the SCN but also in the peripheral tissues indicates the importance of the peripheral clocks at various levels. For example, the plasminogen activator inhibitor-1 (PAI-1) may be an important output gene of the peripheral clocks (Maemura et al. 2007). A screening using cDNA microarray identified more than 20 target genes of the peripheral clocks including transcription factors and membrane receptors. These genes were upregulated by CLOCK/BMAL (Maemura et al. 2007).

Moreover, studies based on genetically engineered mice indicated the involvement of these peripheral clocks in the development of cardiovascular disorders (Takeda and Maemura 2011). The loss of synchronization in both inter- and intraorgans caused by the alterations in the peripheral circadian clocks may contribute to higher cardiovascular risks and cardiometabolic syndrome (Young and Bray 2007). In addition, the inconsistency in the phases between the central and peripheral clocks may have crucial roles in disease development and progression (Takeda and Maemura 2011). In the next section, the roles of circadian patterns in CVDs will be discussed in more details.

5.2 Circadian Patterns and Potential Biomarkers for Cardiovascular Diseases

5.2.1 The Circadian Patterns of Cardiovascular Events

The maintenance of the normal circadian time structure at various levels is critical for promoting cardiovascular health, including the cardiomyocyte level and the organ level (Portaluppi et al. 2012). For example, the feature of morning peak has been observed in arterial pressure, heart rate, and vascular tone (Shaw and Tofler 2009). Such variation may increase plaque rupture and platelet reactivity. Associated with various cardiovascular morbidity and mortality, arterial hypertension may be related to the altered circadian patterns of BP.

Epidemiological studies have confirmed that adverse cardiovascular events often peak in the morning between 6 a.m. and noon (Morris et al. 2012). Such events include myocardial ischemia, acute myocardial infarct, and sudden cardiac death (Portaluppi et al. 2012). In addition, certain day–night patterns have been observed in supraventricular and ventricular cardiac arrhythmias of different types (Portaluppi et al. 2012). Higher frequencies during the day rather than night are the common properties of atrial arrhythmias with atrial fibrillation, flutter, premature beats, and tachycardias. Ventricular fibrillation and ventricular premature beats occur more often in the morning and during the daytime, respectively (Portaluppi et al. 2012).

Adverse cardiovascular events are associated with the high levels of certain bio-markers including cortisol, nonlinear dynamic heart rate, and platelet activation (Morris et al. 2012). These biomarkers are modulated by the internal circadian timing systems. The circadian variations in the hemostatic system such as increased platelet aggregation and coagulation may elevate the incidence of cardiovascular events in the morning (Chrusciel et al. 2009).

Furthermore, the circadian systems may interact with external stressors such as physical activities, resulting in the morning peak in adverse cardiovascular events (Morris et al. 2012). Studies have shown that the circadian systems regulate various cardiovascular risk markers not only at rest, but also during exercise. Such interactions have been considered responsible for the diurnal patterns of many cardiovascular disorders (Scheer et al. 2010).

5.2.2 The Circadian Link Between Cardiovascular and Other Disorders

In addition to the cardiovascular function itself, circadian variations may have a critical role in aging-associated disorders. In older adults, disturbed circadian rest/activity rhythms (RARs) have been associated with higher risks for CVDs-related mortality. For instance, in a study of 2968 men aged 67 years and older, lower levels of circadian robustness such as lower amplitude were related to higher risks for CVDs especially coronary heart disease (CHD), stroke, and peripheral vascular disease (PVD) (Paudel et al. 2011). In another analysis of a total of 874,495 elderly patients, a remarkable peak in the morning was observed in the occurrence of CVDs, cardiopulmonary arrest (CPA), and heat stroke (Kawakami et al. 2008). In a study of 559 elderly patients in India, the circadian variability was observed in acute coronary events as a feature among the aging patients in comparison with their younger counterparts (Bhalla et al. 2006).

Furthermore, circadian rhythms may be one of the key connections linking cardiovascular events with other diseases such as chronic kidney disease (CKD). For example, the 'non-dipper' pattern has been observed in patients with the salt-sensitive type of hypertension or CKD (Kimura et al. 2010). Among these patients, the nighttime BP does not drop as in healthy people, showing an abnormal pattern of BP rhythms. Such nocturnal hypertension and the non-dipper pattern are associated with both cardiovascular events and renal damages at the same time, serving as the pathogenesis of the cardio-renal link. In addition, salt sensitivity and excess salt intake can cause higher glomerular capillary pressure, leading to glomerular sclerosis and finally renal failure (Kimura et al. 2010). The correlations between salt sensitivity and abnormal circadian BP rhythms may provide the basis for understanding the linkages between cardiovascular and renal complications.

5.2.3 *Shift Work, Chronic Circadian Disruptions, and Cardiovascular Risks*

For a systemic understanding of the circadian–cardiovascular interactions, life style may also have a key role. The loss of the synchronization between the internal rhythms and external stimuli may lead to damages in the cardiovascular organs. Alterations in sleep patterns and the loss of circadian rhythmicity have been considered risk factors for CVDs. The connections among shift work, metabolic risks, and CVDs have been well established, representing the pathways from circadian stress to morbidity (Oishi and Ohkura 2013; also see Fig. 5.1).

The misalignment between the behavioral cycles and the circadian rhythms has been associated with both cardiovascular and metabolic diseases among the population of shift workers (Morris et al. 2012). Such connections may be explained via interrelated psychosocial, behavioral, and physiological processes. The psychosocial changes include higher difficulties in controlling working and resting hours, lower work-life balance, and inadequate recovery after work (Puttonen et al. 2010). The behavioral factors include the occurrence of weight gain and smoking. Physiologically, shift work and circadian misalignment may cause altered lipid and glucose metabolism, higher levels of inflammation, and higher risks for atherosclerosis (Morris et al. 2012; Puttonen et al. 2010). These changes have also been associated with obesity and diabetes.

At the cellular level, the loss of the synchronization between the endogenous clocks and behavioral cycles may affect the metabolic and proinflammatory pathways with higher cardiovascular risks (Machado and Koike 2014). For example, atherosclerosis is a chronic inflammatory state with increased cellular proliferation of vascular smooth muscle cells and endothelial cells that may result in myocardial infarction and stroke.

At the molecular level, shift work and chronic circadian disruptions may induce higher expression of the plasminogen activator inhibitor-1 (PAI-1), a crucial regulator of fibrinolysis (Oishi and Ohkura 2013). Such change may lead to higher risks of CVDs. The shift work schedule has also been related to higher plasma resistin levels, which may be associated with the pathogenesis of early metabolic syndrome (Burgueño et al. 2010).

5.3 Chronotherapy for Cardiovascular Diseases

The understanding of the circadian–cardiovascular interactions on the basis of systems biology refers to the importance of the temporal factors in the prevention and treatment of CVDs. The circadian rhythms may have a fundamental role in the pharmacokinetics and pharmacodynamics of drugs for CVDs and other disorders (Portaluppi et al. 2012). Chronotherapy has been recognized as an effective method for promoting the efficacy and decreasing the adverse reactions of these drugs.

Ambulatory BP monitoring may be especially useful for the evaluation of the circadian pharmacodynamics of antihypertensive drugs. For instance, a greater level of BP decrease was observed with telmisartan for the last 6 h of the 24-h dosing interval when it was compared with valsartan (Giles 2005). In another example, a study of 2156 hypertensive patients showed that bedtime chronotherapy had better BP control effects compared to the conventional upon-waking treatment (Hermida et al. 2010). Such strategy also resulted in significantly lower morbidity and mortality in CVDs.

These studies address the remarkable impacts of the treatment time for CVDs. With further understanding of the oscillatory patterns in cardiovascular tissues and cells, novel chronotherapeutic approaches may be developed in accordance with the central and peripheral clocks (Takeda and Maemura 2011). A more detailed discussion on chronotherapy for CVDs will be provided in Chap. 7.

References

Agra RM, Fernández-Trasancos Á, Sierra J, González-Juanatey JR, Eiras S (2014) Differential association of S100A9, an inflammatory marker, and p53, a cell cycle marker, expression with epicardial adipocyte size in patients with cardiovascular disease. Inflammation 37:1504–1512

Bhalla A, Sachdev A, Lehl SS, Singh R, D'Cruz S (2006) Ageing and circadian variation in cardiovascular events. Singapore Med J 47:305–308

Bicknell KA, Brooks G (2008) Reprogramming the cell cycle machinery to treat cardiovascular disease. Curr Opin Pharmacol 8:193–201

Boehm M, Nabel EG (2003) The cell cycle and cardiovascular diseases. Prog Cell Cycle Res 5:19–30

Burgueño A, Gemma C, Gianotti TF, Sookoian S, Pirola CJ (2010) Increased levels of resistin in rotating shift workers: a potential mediator of cardiovascular risk associated with circadian misalignment. Atherosclerosis 210:625–629

Chrusciel P, Goch A, Banach M, Mikhailidis DP, Rysz J, Goch JH (2009) Circadian changes in the hemostatic system in healthy men and patients with cardiovascular diseases. Med Sci Monit 15:RA203–RA208

Czupryniak L, Młynarski W, Pawłowski M, Saryusz-Wolska M, Borkowska A, Klich I, Bodalski J, Loba J (2008) Circadian blood pressure variation in normotensive type 2 diabetes patients and angiotensin converting enzyme polymorphism. Diabetes Res Clin Pract 80:386–391

DBBR (2015) The database of biological rhythms. http://pharmtao.com/health/biological-rhythms-database/. Accessed 1 June 2015

Dominguez-Rodriguez A, Abreu-Gonzalez P, Sanchez-Sanchez JJ, Kaski JC, Reiter RJ (2010) Melatonin and circadian biology in human cardiovascular disease. J Pineal Res 49:14–22

Duez H, Staels B (2008) The nuclear receptors Rev-erbs and RORs integrate circadian rhythms and metabolism. Diab Vasc Dis Res 5:82–88

Fallo F, Dalla Pozza A, Sonino N, Federspil G, Ermani M, Baroselli S, Catena C, Soardo G, Carretta R, Belgrado D et al (2008) Nonalcoholic fatty liver disease, adiponectin and insulin resistance in dipper and nondipper essential hypertensive patients. J Hypertens 26:2191–2197

Giles T (2005) Relevance of blood pressure variation in the circadian onset of cardiovascular events. J Hypertens Suppl 23:S35–S39

Golubnitschaja O (2007) Cell cycle checkpoints: the role and evaluation for early diagnosis of senescence, cardiovascular, cancer, and neurodegenerative diseases. Amino Acids 32:359–371

Herichová I, Šoltésová D, Szántóová K, Mravec B, Neupauerová D, Veselá A, Zeman M (2013) Effect of angiotensin II on rhythmic per2 expression in the suprachiasmatic nucleus and heart and daily rhythm of activity in Wistar rats. Regul Pept 186:49–56

Hermida RC, Ayala DE, Mojón A, Fernández JR (2010) Influence of circadian time of hypertension treatment on cardiovascular risk: results of the MAPEC study. Chronobiol Int 27:1629–1651

Hermida RC, Ayala DE, Fernández JR, Portaluppi F, Fabbian F, Smolensky MH (2011) Circadian rhythms in blood pressure regulation and optimization of hypertension treatment with ACE inhibitor and ARB medications. Am J Hypertens 24:383–391

Kawakami C, Ohshige K, Tochikubo O (2008) Circadian variation in cardiovascular emergencies among the elderly. Clin Exp Hypertens 30:23–31

Kimura G, Dohi Y, Fukuda M (2010) Salt sensitivity and circadian rhythm of blood pressure: the keys to connect CKD with cardiovascular events. Hypertens Res 33:515–520

Kohsaka A, Waki H, Cui H, Gouraud SS, Maeda M (2012) Integration of metabolic and cardiovascular diurnal rhythms by circadian clock. Endocr J 59:447–456

Kulah E, Dursun A, Aktunc E, Acikgoz S, Aydin M, Can M, Dursun A (2007) Effects of angiotensin-converting enzyme gene polymorphism and serum vitamin D levels on ambulatory blood pressure measurement and left ventricular mass in Turkish hypertensive population. Blood Press Monit 12:207–213

Lecarpentier Y, Claes V, Hébert J-L (2010) PPARs, cardiovascular metabolism, and function: near- or far-from-equilibrium pathways. PPAR Res 2010:783273

Lecarpentier Y, Claes V, Duthoit G, Hébert J-L (2014) Circadian rhythms, Wnt/beta-catenin pathway and PPAR alpha/gamma profiles in diseases with primary or secondary cardiac dysfunction. Front Physiol 5:429

Ma H, Zhong W, Jiang Y, Fontaine C, Li S, Fu J, Olkkonen VM, Staels B, Yan D (2013) Increased atherosclerotic lesions in LDL receptor deficient mice with hematopoietic nuclear receptor Rev-erbα knock-down. J Am Heart Assoc 2:e000235

Machado RM, Koike MK (2014) Circadian rhythm, sleep pattern, and metabolic consequences: an overview on cardiovascular risk factors. Horm Mol Biol Clin Invest 18:47–52

Maemura K, Takeda N, Nagai R (2007) Circadian rhythms in the CNS and peripheral clock disorders: role of the biological clock in cardiovascular diseases. J Pharmacol Sci 103:134–138

Martino TA, Tata N, Belsham DD, Chalmers J, Straume M, Lee P, Pribiag H, Khaper N, Liu PP, Dawood F et al (2007) Disturbed diurnal rhythm alters gene expression and exacerbates cardiovascular disease with rescue by resynchronization. Hypertension 49:1104–1113

Möller-Levet CS, Archer SN, Bucca G, Laing EE, Slak A, Kabiljo R, Lo JCY, Santhi N, von Schantz M, Smith CP et al (2013) Effects of insufficient sleep on circadian rhythmicity and expression amplitude of the human blood transcriptome. Proc Natl Acad Sci U S A 110:E1132–E1141

Morris CJ, Yang JN, Scheer FAJL (2012) The impact of the circadian timing system on cardiovascular and metabolic function. Prog Brain Res 199:337–358

Mousa TM, Schiller AM, Zucker IH (2014) Disruption of cardiovascular circadian rhythms in mice post myocardial infarction: relationship with central angiotensin II receptor expression. Physiol Rep 2:e12210

Nakagawa Y, Kishida K, Kihara S, Sonoda M, Hirata A, Yasui A, Nishizawa H, Nakamura T, Yoshida R, Shimomura I et al (2008) Nocturnal reduction in circulating adiponectin concentrations related to hypoxic stress in severe obstructive sleep apnea-hypopnea syndrome. Am J Physiol Endocrinol Metab 294:E778–E784

Ohmori M, Fujimura A (2005) ACE inhibitors and chronotherapy. Clin Exp Hypertens 27:179–185

Oishi K, Ohkura N (2013) Chronic circadian clock disruption induces expression of the cardiovascular risk factor plasminogen activator inhibitor-1 in mice. Blood Coagul Fibrinolysis 24:106–108

Oishi K, Uchida D, Ohkura N, Horie S (2010) PPARα deficiency augments a ketogenic diet-induced circadian PAI-1 expression possibly through PPARγ activation in the liver. Biochem Biophys Res Commun 401:313–318

Paudel ML, Taylor BC, Ancoli-Israel S, Stone KL, Tranah G, Redline S, Barrett-Connor E, Stefanick ML, Ensrud KE (2011) Rest/activity rhythms and cardiovascular disease in older men. Chronobiol Int 28:258–266

Petkova SB, Ashton A, Bouzahzah B, Huang H, Pestell RG, Tanowitz HB (2000) Cell cycle molecules and diseases of the cardiovascular system. Front Biosci 5:D452–D460

Portaluppi F, Tiseo R, Smolensky MH, Hermida RC, Ayala DE, Fabbian F (2012) Circadian rhythms and cardiovascular health. Sleep Med Rev 16:151–166

Puttonen S, Härmä M, Hublin C (2010) Shift work and cardiovascular disease – pathways from circadian stress to morbidity. Scand J Work Environ Health 36:96–108

Scheer FAJL, Hu K, Evoniuk H, Kelly EE, Malhotra A, Hilton MF, Shea SA (2010) Impact of the human circadian system, exercise, and their interaction on cardiovascular function. Proc Natl Acad Sci U S A 107:20541–20546

Schroeder A, Loh DH, Jordan MC, Roos KP, Colwell CS (2011) Circadian regulation of cardiovascular function: a role for vasoactive intestinal peptide. Am J Physiol Heart Circ Physiol 300:H241–H250

Shaw E, Tofler GH (2009) Circadian rhythm and cardiovascular disease. Curr Atheroscler Rep 11:289–295

Solt LA, Wang Y, Banerjee S, Hughes T, Kojetin DJ, Lundasen T, Shin Y, Liu J, Cameron MD, Noel R et al (2012) Regulation of circadian behaviour and metabolism by synthetic REV-ERB agonists. Nature 485:62–68

Spiering W, Zwaan IM, Kroon AA, de Leeuw PW (2005) Genetic influences on 24 h blood pressure profiles in a hypertensive population: role of the angiotensin-converting enzyme insertion/deletion and angiotensin II type 1 receptor A1166C gene polymorphisms. Blood Press Monit 10:135–141

Takeda N, Maemura K (2011) Circadian clock and cardiovascular disease. J Cardiol 57:249–256

Vukolic A, Antic V, Van Vliet BN, Yang Z, Albrecht U, Montani J-P (2010) Role of mutation of the circadian clock gene Per2 in cardiovascular circadian rhythms. Am J Physiol Regul Integr Comp Physiol 298:R627–R634

Wang J, Yin L, Lazar MA (2006) The orphan nuclear receptor Rev-erb alpha regulates circadian expression of plasminogen activator inhibitor type 1. J Biol Chem 281:33842–33848

Wang N, Yang G, Jia Z, Zhang H, Aoyagi T, Soodvilai S, Symons JD, Schnermann JB, Gonzalez FJ, Litwin SE et al (2008) Vascular PPARgamma controls circadian variation in blood pressure and heart rate through Bmal1. Cell Metab 8:482–491

Yang G, Jia Z, Aoyagi T, McClain D, Mortensen RM, Yang T (2012) Systemic PPARγ deletion impairs circadian rhythms of behavior and metabolism. PLoS One 7:e38117

Young ME, Bray MS (2007) Potential role for peripheral circadian clock dyssynchrony in the pathogenesis of cardiovascular dysfunction. Sleep Med 8:656–667

Chapter 6
Circadian Rhythms and Cellular Networks in Cancer

6.1 The Circadian Clocks and Cancer: Potential Molecular and Cellular Biomarkers

6.1.1 The Circadian–Cancer Correlations: A Systems Biology Perspective

Studies based on systems biology would allow for an integrative understanding of the circadian–cancer correlations. At the molecular level, the circadian clocks are driven by the feedback loops of proteins with periodic activation and repression (Soták et al. 2014; also see Chap. 2). At the cellular level, the clocks regulate the daily activities in cell cycles, proliferation, metabolism, DNA repair, as well as apoptosis. Peripheral autonomous circadian rhythms in various tissues are synchronized via the master regulator in the suprachiasmatic nucleus (SCN) (Kelleher et al. 2014). The coordination of the central and peripheral clocks is associated with rhythmic and tissue-specific gene expressions in the responses to cyclic environmental changes (Fu and Kettner 2013). At the organismal level, the clocks are correlated with energy homeostasis, neuroendocrine and immune functions, as well as various physical activities (Kettner et al. 2014). Many causes may lead to the disturbance of the endogenous circadian homeostasis, such as alterations in lifestyle in the current industrialized society (Kettner et al. 2014; Soták et al. 2014).

As an important risk factor, the disruption of the clock functions at different levels may contribute to the development and progression of cancer. Molecular, cellular, and clinical studies have identified biological and behavioral factors associated with circadian patterns as potential biomarkers (Eismann et al. 2010). Further understanding of the mechanisms underlying the clock-modulated tumor suppression may promote personalized prevention and treatment of cancer. For example, various studies have confirmed that those with strong and robust circadian rhythms have less signs of tumor progression as measured by the biomarkers including VEGF and TGFβ (Cash et al. 2015). On the other hand, among those with active and

© Springer International Publishing Switzerland 2015
Q. Yan, *Cellular Rhythms and Networks: Implications for Systems Medicine*,
SpringerBriefs in Cell Biology, DOI 10.1007/978-3-319-22819-8_6

aggressive breast tumors, circadian disruptions and higher levels of cortisol awakening response (CAR) may contribute to tumor development and progression.

Alterations in the clock genes such as mutations and epigenetic silencing have been related to higher risks for cancer (Kettner et al. 2014). These genes include CLOCK, BMAL1 and NPAS2 that trigger the transcriptions of PER1, PER2, as well as CRY1 and CRY2 (Kelleher et al. 2014). Such changes in the circadian genes may also contribute to aging-associated phenotypes (Yu and Weaver 2011). Specifically, studies in mice models have indicated that mutations in CRY1 and CRY2 may lead to circadian disruptions and accelerated growth of implanted tumors (Kelleher et al. 2014). Lower levels of CRY2 have been observed in breast cancer tissues (Mao et al. 2015). In addition, CRY2 may be involved in the breast cancer progression and prognosis via altered methylation in cancer-associated pathways.

The circadian clocks are transcriptional systems controlled by epigenetic factors including histone modifications and DNA methylation, which may connect the clocks with cancer cells initiation and progression (Masri et al. 2015; Joska et al. 2014). Abnormal DNA methylation has been suggested as a potential biomarker for cancer. For example, one of the core circadian genes ARNTL (also known as BMAL1) is epigenetically silenced in ovarian cancer (Yeh et al. 2014). On the other hand, the overexpression of ARNTL (BMAL1) has been found to suppress cell growth, promote the chemosensitivity of cisplatin, and reestablish the rhythmic processes of c-MYC in cancer cells. These findings suggest that normal ARNTL (BMAL1) may have tumor suppressor functions (Yeh et al. 2014). In the next section, more examples and detailed discussions of circadian-associated genes and pathways will be provided.

6.1.2 Examples of Circadian-Associated Molecular Factors and Cellular Networks

Table 6.1 lists some examples of circadian-associated molecular factors and cellular networks that may have important roles in cancer and chronotherapy. Some of these factors are discussed in details below. A more complete list can be found in the Database of Biological Rhythms (DBBR 2015).

6.1.2.1 TIMELESS and Associated Networks

The circadian gene TIMELESS (TIM) has been found vital for replication protection and genomic stability (Yoshida et al. 2013). It may also have critical roles in human tumorigenesis and malignancies. Studies using expression profiling have suggested it as a potential biomarker for cancer risks and prognosis with implications as a therapeutic target (Yoshida et al. 2013). For example, in breast cancer and lung cancer samples, the overexpression of TIMELESS has been related to more advanced tumor stages and poorer prognostic outcomes (Mao et al. 2013; Yoshida et al. 2013). On the other hand, its knockdown inhibited proliferation and clonogenic growth, and promoted apoptosis in cancer cells.

Table 6.1 Examples of circadian-associated genes and pathways involved in cancer

Genes	Changes/interactions	Roles in cancer	References
EGFR	Circadian inhibition of signaling by glucocorticoids	Chronotherapy	Lauriola et al. (2014)
	Inhibition on hypothalamic signaling of rhythmic behaviors	A symptom cluster in metastatic colorectal cancer	Rich (2007)
	Melatonin, EGFR/mitogen-activated protein kinase (MAPK) pathways	Efficacy and toxicity of chemo- and radiotherapy	Blask et al. (2002)
NFκB	Daily variations in the intensity, CLOCK	Immune responses in embryonic fibroblasts and primary hepatocytes	Spengler et al. (2012)
	CRY, p53, TNFα	Circadian regulation of apoptosis	Lee and Sancar (2011)
	CRY, cAMP, PKA	Chronic inflammation	Narasimamurthy et al. (2012)
TIMELESS	Overexpression	Cancer cell proliferation, breast cancer prognosis	Mao et al. (2013)
	Higher expression	A diagnostic and prognostic marker in lung cancer	Yoshida et al. (2013)
	Overexpression, SNPs, promoter hypomethylation	Risks and prognosis of breast cancer	Fu et al. (2012)
	DNA damage pathways	A potential drug target	Yang et al. (2010)
TP53	Circadian patterns of the erythemal response	Altered DNA repair and apoptosis	Gaddameedhi et al. (2015)
	Cell cycle genes, c-MYC, PER2	Altered apoptosis	Iwamoto et al. (2014)
	Circadian accumulation, ATF4	Chronotherapy	Horiguchi et al. (2013)
	CRY, TNFα-initiated apoptosis, NFκB pathways	Delayed onset of tumorigenesis	Lee and Sancar (2011)
	BMAL1, PER1, PER2, PER3, WEE1	Tumor development, drug responses	Zeng et al. (2010)
	ARNTL (BMAL1), p21(CIP1)	Altered apoptosis	Mullenders et al. (2009)
	CRY	Protection from the early onset of cancer, apoptosis	Ozturk et al. (2009)
	c-MYC	Cancer growth	Filipski et al. (2006)

The circadian clocks and cell cycle are general regulatory systems with physiological effects on normal cells and pathological impacts on cancer cells. As a regulator of both circadian and cell cycles, TIMELESS may serve as the key connection between these two systems. Specifically, it has been found necessary for the ATR-mediated CHK1 activation and S-phase arrest, as well as for ATM-mediated CHK2 activation and G2/M checkpoint regulation (Yang et al. 2010). It is also involved in the signaling of doxorubicin-mediated DNA double strand breaks. Its depletion may promote the doxorubicin-mediated cytotoxicity in cancer cells (Yang et al. 2010). These mechanisms indicate that it can become a potential anticancer target.

In addition, TIMELESS may serve as the link between circadian patterns and breast cancer risks with genetic and epigenetic correlations. It may be involved in various pathways associated with cancer including the hormone regulation network. Genetic and epigenetic association studies have identified the relationships between two SNPs (rs2291738 and rs7302060) in the TIMELESS gene and lower risks of breast cancer among ER- or PR-positive patients (Fu et al. 2012). Links have also been established between breast cancer in the stages II, III, and IV and TIMELESS promoter hypomethylation (Fu et al. 2012). These findings confirm its role as a potential biomarker for breast cancer.

6.1.2.2   P53 and Associated Networks

The p53 tumor suppressor gene has a key role in cancer as it is mutated in about 50 % of human cancers (Mullenders et al. 2009). In most of the cancer cases the p53 pathways are functionally inactivated. The elucidation of the mechanisms underlying the insensitivity to p53 activation among tumor cells would be necessary for better treatment of cancer. For instance, studies using mouse models showed that the circadian clocks may be pivotal in the regulation of sunburn apoptosis and erythema (Kettner et al. 2014). In normal conditions, the levels of p53 that regulates the process of sunburn apoptosis are maximally induced in the morning. However, the mutational inactivation of p53 may be associated with the higher susceptibility to skin cancer induction after chronic irradiation in the morning (Kettner et al. 2014). Furthermore, studies of tumors in jet lagged mice showed that circadian disruptions may lead to lower levels of p53 with the overexpression of c-MYC, the two effects that may enhance cancer growth (Filipski et al. 2006).

In addition, the circadian clock gene ARNTL (also known as BMAL1) may be involved in the p53 tumor suppressor pathway. ARNTL (BMAL1) may serve as the linkage connecting circadian rhythms with cell cycle and cancer (Mullenders et al. 2009). The loss of ARNTL (BMAL1) has been found to decrease the expression of PER1, PER2, PER3, WEE1, as well as p53 (Zeng et al. 2010). Cells with lower levels of ARNTL (BMAL1) may lose the ability to arrest mediated by the inability to trigger the p53 target genes (Mullenders et al. 2009). The interrelationship between ARNTL (BMAL1) and p53 further supports the connection between circadian rhythms and cancer.

Most chemotherapeutic drugs may cause DNA damages and cell death via the p53 associated pathways. However, the sensitivity of cancer cells to chemotherapeutic agents may alternate in accordance with the circadian oscillations. Specifically, the transcription factor and circadian gene ATF4 may be involved in the circadian accrual of the p53 protein in tumor cells via the regulation of the p19ARF-MDM2 pathway (Horiguchi et al. 2013). This circadian rhythmic regulation indicates that the cancer cell sensitivity to chemotherapeutic agents may be promoted during certain time-of-day with the accumulation of the p53 protein.

6.2 The Circadian–Cell Cycle Interactions and Cancer

As mentioned earlier, circadian rhythms and cell cycle progression are the two most basic biological rhythmic activities with bidirectional interactions. While circadian oscillations support the temporal regulation in the 24-h cycle, the cell cycle is involved in cellular growth and division. Because the circadian systems are tightly interconnected with cell proliferation and apoptosis, the disruptions of the circadian–cell cycle interactions may be associated with cancer development and progression (Soták et al. 2014).

Figure 6.1 shows a schema of some typical interactions between circadian pathways and cellular processes associated with cancer, including cell cycle checkpoints, cell proliferation and apoptosis, DNA repair, metabolism, and stress responses. Each process may have complex and multiple pathways involved. Specifically, the positive/negative feedback loops of proteins with periodic activation and inhibition are involved in the regulation of both cell cycle progression and circadian rhythms (Soták et al. 2014). The elucidation of these interrelationships may be helpful for the development of novel preventive and therapeutic strategies for cancer.

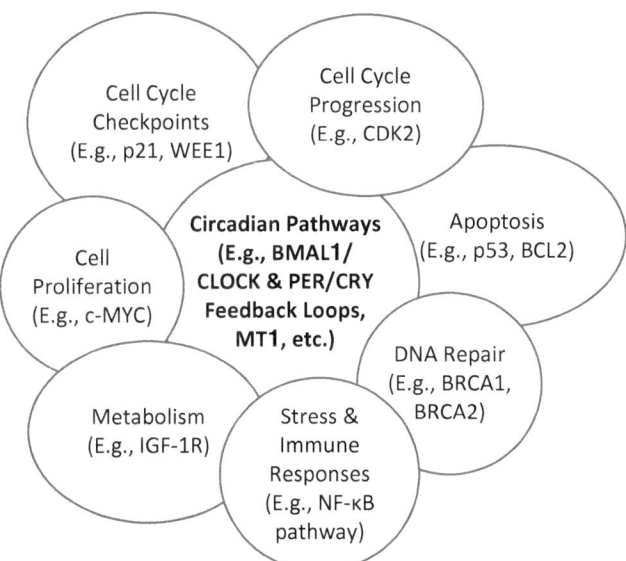

Fig. 6.1 The interactions between circadian pathways and cellular processes in cancer

The expression of cell cycle genes can be influenced by various factors. For example, a microarray analysis showed that light pulses administered during the circadian dark phase may affect the transcription levels of the genes directly regulating the cell cycle and result in cell cycle arrest (Ben-Shlomo and Kyriacou 2010). In the meantime, alterations were also observed in the transcription levels of the genes involved in tumorigenesis and metastasis. In addition, abnormal expression of the circadian genes may result in genomic instability and enhanced cellular proliferation with impacts on the cell cycle genes including c-MYC, WEE1, cyclin D and p21 (Kelleher et al. 2014). For example, altered PER2 may cause the upregulation of c-MYC with a higher tumor incidence (Kelleher et al. 2014).

In a study among patients with chronic lymphocytic leukemia (CLL), downregulated expressions of the circadian genes BMAL1, PER1, and PER2 were observed (Rana et al. 2014). In the meantime, the expressions of the cell cycle genes MYC and cyclin D1 were upregulated. Furthermore, the expressions of these genes were more aberrant among shift-workers than non-shift-workers in the CLL patients. Decreased serum melatonin levels were also observed, and even lower among shift-workers (Rana et al. 2014).

Such observations demonstrate the complex interactions among life style, circadian patterns, and cell cycles. The abnormal expressions of the clock genes may result in the alterations of their downstream targets associated with cell proliferation and apoptosis in cancer patients. In addition, life style factors such as shift work and low melatonin levels may be involved in the circadian disruptions and contribute to the disease development (Rana et al. 2014).

In another study, the mutation S662G in the circadian protein PER2 was found to have a key role in cell cycle progression and tumorigenesis (Gu et al. 2012). This mutation has also been related to familial advanced sleep phase syndrome. In cancer therapy, this mutation may promote the resistance to X-ray-induced apoptosis and higher levels of E1A- and RAS-mediated oncogenic transformation (Gu et al. 2012). Furthermore, the analysis of the clock-regulated cell cycle genes showed that the mutation may lead to alterations in the relative phases between p21 and cyclin D expression profiles in mouse embryonic fibroblasts (Gu et al. 2012). Such PER2-associated phase change of p21 has been suggested as an essential mechanism underlying the cell cycle progression and therapeutic resistance in cancer.

The understanding of the circadian–cell cycle interactions may also contribute to the optimization of drug delivery via chronotherapy. For example, the anticancer drug 5-fluorouracil (5-FU) kills cells in the S phase. A study using a cell cycle automaton model found that the cytotoxicity of 5-FU may reach the lowest level for the circadian delivery peaking at 4 a.m., and the highest level for the continuous infusion or the circadian pattern peaking at 4 p.m. (Altinok et al. 2007). The automaton model illustrated the transitions through the continuous stages of the cell cycle and the entrainment of the cell cycle by the circadian oscillation. Considering different phases of the cell cycle, such models based on cell cycle automaton may be useful for the prediction of the cytotoxic effect of anticancer drugs (Altinok et al. 2007). More discussions of cancer chronotherapy can be found in Chap. 7.

6.3 Alterations in Circadian Patterns in Various Cell Types

As the circadian systems regulate cellular metabolism and proliferation, the disturbance may accelerate cancer development and progression in different tissues and cell types. Clinical studies have indicated that circadian disruptions are associated with suppressed cellular immunity and increased inflammatory responses (Eismann et al. 2010). Such changes may lead to the promotion of tumor growth, angiogenesis, and metastasis.

In addition, tissue culture studies have shown that biological or behavioral factors may cause alterations in the circadian pathways and faster growth of tumor cells. In certain cancers such as breast cancer, the involvement of psychoneuroendocrine and psychoneuroimmune pathways has been emphasized (Eismann et al. 2010). In kidney cancer, close interrelationships have been identified between the hypoxic response pathways and circadian pathways (Mazzoccoli et al. 2014). Dysfunctions in the circadian genes and abnormal hypoxia reactions have been associated with the development of carcinogenesis in kidney cancer patients.

Being the first line of guard against environmental and stress impacts, the skin tissues are influenced by diurnal oscillations in solar ultraviolet (UV) radiation and temperature. The functional circadian clocks have been identified in most of the skin cell types. Connections have been established between the circadian clocks and skin cancers, especially the modulation of UVB-induced DNA damages (Plikus et al. 2015). Because the skin also provides the protection against microorganisms, the circadian clocks in the skin tissues may be closely associated with immune functions. For example, the links have been found between the circadian clocks and psoriasis, an immune-mediated skin disease (Plikus et al. 2015). In addition to diurnal rhythms, many animal models have obvious seasonal hair molt cycles, indicating the importance of the clocks in seasonal organismal behaviors.

Epidemiological analyses such as "The Nurses' Health Study" have suggested that women working rotational night shifts may have higher risks for breast cancer (Kelleher et al. 2014). In addition to cancer, the circadian disruption may cause jet-lag-like symptoms in short terms and lead to obesity, metabolic syndrome, type 2 diabetes, as well as cardiovascular disease in the long run (Haus and Smolensky 2013). The mechanisms underlying the connection between shift work and cancer development include circadian disruptions, melatonin suppression, immune dysfunction, and metabolic alterations. The increase of proinflammatory reactive oxygen species may also have a role (Haus and Smolensky 2013).

At the molecular level, the loss of circadian genes such as CRY2 and PER3 has been related to tumor progression and worse prognosis in breast cancer (Cadenas et al. 2014). Among shift workers, the gene-shift work interactions were observed for the melatonin signaling gene MTNR1B with the key transcription factors NPAS2 and ARNTL (also known as BMAL1) (Rabstein et al. 2014). Multivariable models involving clock gene polymorphisms have been suggested to affect breast cancer risks, chrono-types, and light sensitivity (Rabstein et al. 2014; Truong et al. 2014). For example, genetic variations in the circadian pathway such as the SNPs in

RORA (rs1482057 and rs12914272) and rs11932595 in CLOCK have been related to breast cancer risks among postmenopausal women (Truong et al. 2014).

In addition, melatonin is involved in not only the control of circadian clocks, but also the integration of molecular, dietary, and metabolic signaling activities associated with breast cancer growth. The anti-proliferative functions of melatonin may be mediated via its interactions with MT(1) melatonin receptors in human breast cancer cell lines (Blask et al. 2011). Melatonin may be involved in a variety of interactions and pathways including the estrogen receptor pathway, especially the suppression of cAMP levels and growth factor pathways associated with cell proliferation (Blask et al. 2011).

Studies have confirmed that dim light at night (LAN) may cause lower melatonin levels, the circadian-mediated host-cancer imbalance, and higher breast cancer risks among shift workers and other individuals with higher exposure to LAN (Blask et al. 2011, 2014). Such alterations may result in hyperglycemia and hyperinsulinemia in the host. Other results include abnormal aerobic glycolysis, lipid signaling and proliferative processes in the tumors (Blask et al. 2014). Further studies would be needed to elucidate the roles of circadian control in various cell proliferation and carcinogenesis to support better strategies in chronotherapy.

References

Altinok A, Lévi F, Goldbeter A (2007) A cell cycle automaton model for probing circadian patterns of anticancer drug delivery. Adv Drug Deliv Rev 59:1036–1053

Blask DE, Sauer LA, Dauchy RT (2002) Melatonin as a chronobiotic/anticancer agent: cellular, biochemical, and molecular mechanisms of action and their implications for circadian-based cancer therapy. Curr Top Med Chem 2:113–132

Blask DE, Hill SM, Dauchy RT, Xiang S, Yuan L, Duplessis T, Mao L, Dauchy E, Sauer LA (2011) Circadian regulation of molecular, dietary, and metabolic signaling mechanisms of human breast cancer growth by the nocturnal melatonin signal and the consequences of its disruption by light at night. J Pineal Res 51:259–269

Blask DE, Dauchy RT, Dauchy EM, Mao L, Hill SM, Greene MW, Belancio VP, Sauer LA, Davidson L (2014) Light exposure at night disrupts host/cancer circadian regulatory dynamics: impact on the Warburg effect, lipid signaling and tumor growth prevention. PLoS One 9:e102776

Cadenas C, van de Sandt L, Edlund K, Lohr M, Hellwig B, Marchan R, Schmidt M, Rahnenführer J, Oster H, Hengstler JG (2014) Loss of circadian clock gene expression is associated with tumor progression in breast cancer. Cell Cycle 13:3282–3291

Cash E, Sephton SE, Chagpar AB, Spiegel D, Rebholz WN, Zimmaro LA, Tillie JM, Dhabhar FS (2015) Circadian disruption and biomarkers of tumor progression in breast cancer patients awaiting surgery. Brain Behav Immun 48:102–114

DBBR (2015) The database of biological rhythms. http://pharmtao.com/health/biological-rhythms-database/. Accessed 1 June 2015

Eismann EA, Lush E, Sephton SE (2010) Circadian effects in cancer-relevant psychoneuroendocrine and immune pathways. Psychoneuroendocrinology 35:963–976

Filipski E, Li XM, Lévi F (2006) Disruption of circadian coordination and malignant growth. Cancer Causes Control 17:509–514

Fu L, Kettner NM (2013) The circadian clock in cancer development and therapy. Prog Mol Biol Transl Sci 119:221–282

Fu A, Leaderer D, Zheng T, Hoffman AE, Stevens RG, Zhu Y (2012) Genetic and epigenetic associations of circadian gene TIMELESS and breast cancer risk. Mol Carcinog 51:923–929

Gaddameedhi S, Selby CP, Kemp MG, Ye R, Sancar A (2015) The circadian clock controls sunburn apoptosis and erythema in mouse skin. J Invest Dermatol 135:1119–1127

Gu X, Xing L, Shi G, Liu Z, Wang X, Qu Z, Wu X, Dong Z, Gao X, Liu G et al (2012) The circadian mutation PER2(S662G) is linked to cell cycle progression and tumorigenesis. Cell Death Differ 19:397–405

Haus EL, Smolensky MH (2013) Shift work and cancer risk: potential mechanistic roles of circadian disruption, light at night, and sleep deprivation. Sleep Med Rev 17:273–284

Horiguchi M, Koyanagi S, Hamdan AM, Kakimoto K, Matsunaga N, Yamashita C, Ohdo S (2013) Rhythmic control of the ARF-MDM2 pathway by ATF4 underlies circadian accumulation of p53 in malignant cells. Cancer Res 73:2639–2649

Iwamoto A, Kawai M, Furuse M, Yasuo S (2014) Effects of chronic jet lag on the central and peripheral circadian clocks in CBA/N mice. Chronobiol Int 31:189–198

Joska TM, Zaman R, Belden WJ (2014) Regulated DNA methylation and the circadian clock: implications in cancer. Biology (Basel) 3:560–577

Kelleher FC, Rao A, Maguire A (2014) Circadian molecular clocks and cancer. Cancer Lett 342:9–18

Kettner NM, Katchy CA, Fu L (2014) Circadian gene variants in cancer. Ann Med 46:208–220

Lauriola M, Enuka Y, Zeisel A, D'Uva G, Roth L, Sharon-Sevilla M, Lindzen M, Sharma K, Nevo N, Feldman M et al (2014) Diurnal suppression of EGFR signalling by glucocorticoids and implications for tumour progression and treatment. Nat Commun 5:5073

Lee JH, Sancar A (2011) Regulation of apoptosis by the circadian clock through NF-kappaB signaling. Proc Natl Acad Sci U S A 108:12036–12041

Mao Y, Fu A, Leaderer D, Zheng T, Chen K, Zhu Y (2013) Potential cancer-related role of circadian gene TIMELESS suggested by expression profiling and in vitro analyses. BMC Cancer 13:498

Mao Y, Fu A, Hoffman AE, Jacobs DI, Jin M, Chen K, Zhu Y (2015) The circadian gene CRY2 is associated with breast cancer aggressiveness possibly via epigenomic modifications. Tumour Biol 36:3533–3539

Masri S, Kinouchi K, Sassone-Corsi P (2015) Circadian clocks, epigenetics, and cancer. Curr Opin Oncol 27:50–56

Mazzoccoli G, De Cata A, Piepoli A, Vinciguerra M (2014) The circadian clock and the hypoxic response pathway in kidney cancer. Tumour Biol 35:1–7

Mullenders J, Fabius AWM, Madiredjo M, Bernards R, Beijersbergen RL (2009) A large scale shRNA barcode screen identifies the circadian clock component ARNTL as putative regulator of the p53 tumor suppressor pathway. PLoS One 4:e4798

Narasimamurthy R, Hatori M, Nayak SK, Liu F, Panda S, Verma IM (2012) Circadian clock protein cryptochrome regulates the expression of proinflammatory cytokines. Proc Natl Acad Sci U S A 109:12662–12667

Ozturk N, Lee JH, Gaddameedhi S, Sancar A (2009) Loss of cryptochrome reduces cancer risk in p53 mutant mice. Proc Natl Acad Sci U S A 106:2841–2846

Plikus MV, Van Spyk EN, Pham K, Geyfman M, Kumar V, Takahashi JS, Andersen B (2015) The circadian clock in skin: implications for adult stem cells, tissue regeneration, cancer, aging, and immunity. J Biol Rhythms 30:163–182

Rabstein S, Harth V, Justenhoven C, Pesch B, Plöttner S, Heinze E, Lotz A, Baisch C, Schiffermann M, Brauch H et al (2014) Polymorphisms in circadian genes, night work and breast cancer: results from the GENICA study. Chronobiol Int 31:1115–1122

Rana S, Munawar M, Shahid A, Malik M, Ullah H, Fatima W, Mohsin S, Mahmood S (2014) Deregulated expression of circadian clock and clock-controlled cell cycle genes in chronic lymphocytic leukemia. Mol Biol Rep 41:95–103

Rich TA (2007) Symptom clusters in cancer patients and their relation to EGFR ligand modulation of the circadian axis. J Support Oncol 5:167–174, discussion 176–177

Ben-Shlomo R, Kyriacou CP (2010) Light pulses administered during the circadian dark phase alter expression of cell cycle associated transcripts in mouse brain. Cancer Genet Cytogenet 197:65–70

Soták M, Sumová A, Pácha J (2014) Cross-talk between the circadian clock and the cell cycle in cancer. Ann Med 46:221–232

Spengler ML, Kuropatwinski KK, Comas M, Gasparian AV, Fedtsova N, Gleiberman AS, Gitlin II, Artemicheva NM, Deluca KA, Gudkov AV et al (2012) Core circadian protein CLOCK is a positive regulator of NF-κB-mediated transcription. Proc Natl Acad Sci U S A 109:E2457–E2465

Truong T, Liquet B, Menegaux F, Plancoulaine S, Laurent-Puig P, Mulot C, Cordina-Duverger E, Sanchez M, Arveux P, Kerbrat P et al (2014) Breast cancer risk, nightwork, and circadian clock gene polymorphisms. Endocr Relat Cancer 21:629–638

Yang X, Wood PA, Hrushesky WJM (2010) Mammalian TIMELESS is required for ATM-dependent CHK2 activation and G2/M checkpoint control. J Biol Chem 285:3030–3034

Yeh C-M, Shay J, Zeng T-C, Chou J-L, Huang TH-M, Lai H-C, Chan MWY (2014) Epigenetic silencing of ARNTL, a circadian gene and potential tumor suppressor in ovarian cancer. Int J Oncol 45:2101–2107

Yoshida K, Sato M, Hase T, Elshazley M, Yamashita R, Usami N, Taniguchi T, Yokoi K, Nakamura S, Kondo M et al (2013) TIMELESS is overexpressed in lung cancer and its expression correlates with poor patient survival. Cancer Sci 104:171–177

Yu EA, Weaver DR (2011) Disrupting the circadian clock: gene-specific effects on aging, cancer, and other phenotypes. Aging (Albany, NY) 3:479–493

Zeng Z-L, Wu M-W, Sun J, Sun Y-L, Cai Y-C, Huang Y-J, Xian L-J (2010) Effects of the biological clock gene Bmal1 on tumour growth and anti-cancer drug activity. J Biochem 148:319–326

Chapter 7
Circadian Biomarkers and Chronotherapy: Implications for Personalized and Systems Medicine

7.1 From Circadian Rhythms to Chronotherapy: The Roles of Systems Biology

7.1.1 A Systemic Framework for Chronobiology-Based Personalized Medicine

As an emerging concept, chronotherapy refers to the adaptation of the timing of drug administration to promote the overall efficacy of the treatment and to minimize the possible side-effects (Kaur et al. 2013). Although still in the conceptualization or early experimental stages, the fast development in chronotherapy is promising for enriching the practice of personalized and systems medicine.

Figure 7.1 provides a systemic framework for chronobiology-based personalized medicine. It shows the pathways of how the circadian systems such as the BMAL1/CLOCK and PER/CRY feedback loops can be applied for personalized diagnosis and treatment. First of all, biomarkers based on chronobiology and systems biology can be used for the establishment of systemic rhythmic profiles toward more precise diagnosis, prognosis, and individualized treatment. For better clinical practice, the temporal factors such as the circadian rhythmicity are essential in the drug absorption, distribution, metabolism, and excretion (ADME). Strategies in chronotherapy can be applied for the optimized timing, amount, and composition of drug administration to promote the efficacy and minimize the toxicity (see Fig. 7.1).

Furthermore, novel drugs based on chronobiology can be designed to target the relevant circadian mechanisms and chrono-biomarkers associated with pathogenesis such as inflammation (see Chaps. 3–6). On the basis of the robust chrono-biomarkers and chronotherapy, personalized medicine may be achieved via the systems and dynamical approaches (see Fig. 7.1 and Chap. 1).

Preliminary screening of novel therapeutics for their potentials in chronotherapy may become an innovative strategy for promoting the quality of medical practice.

© Springer International Publishing Switzerland 2015 71
Q. Yan, *Cellular Rhythms and Networks: Implications for Systems Medicine*,
SpringerBriefs in Cell Biology, DOI 10.1007/978-3-319-22819-8_7

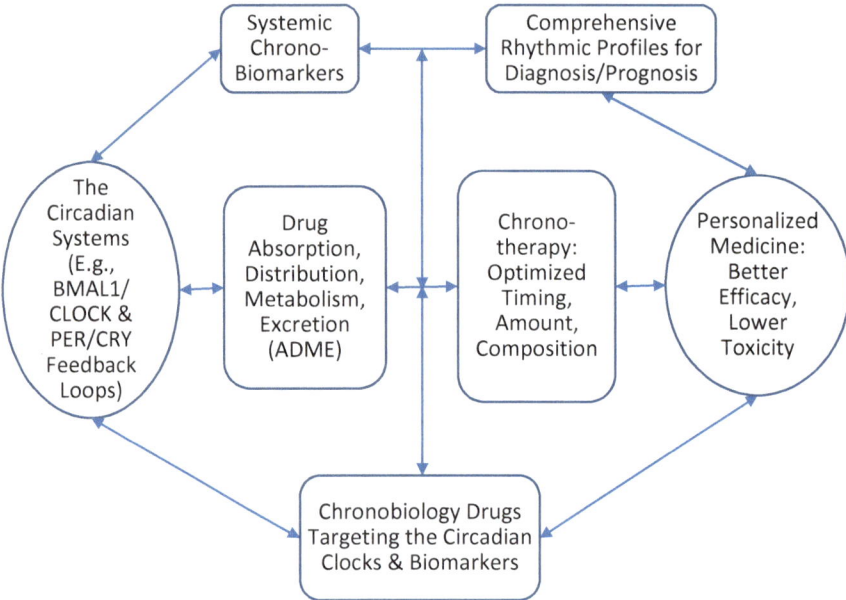

Fig. 7.1   A systemic framework for chronobiology-based personalized medicine

Together with the development in systems biology and pharmacogenomics, studies in chronotherapy should embrace systemic factors including molecular, cellular, age, gender, and lifestyle associated differences.

7.1.2   Systems Biology Approaches in Chronotherapy

With the oscillation of the circadian system, various cellular functions from cell cycle to proliferation, from DNA repair to apoptosis, from pharmacology to toxicology also oscillate hierarchically by the time-of-day (Lévi et al. 2007, 2010). At the organ and tissue levels, the suprachiasmatic nucleus (SCN) of the hypothalamus hosts the "master clock" coordinating physiological activities and behaviors (Dallmann et al. 2014). Other brain areas and body tissues are also involved in the regulation of various oscillations (see Chaps. 2–6). These oscillations influence the drug absorption, metabolism, transportation, and detoxification. The analysis of these processes requests the understanding in chronopharmacology and specifically, circadian pharmacokinetics (PK) and pharmacodynamics (PD) (Dallmann et al. 2014).

In addition to the temporal elements such as the circadian rhythms, the optimal chronotherapy scheduling can be influenced by other factors including genetic settings, cell cycle, gender, and lifestyle. For example, a large clinical trial indicated that gender was associated with the survival outcome and chronotherapeutic delivery among cancer patients (Lévi et al. 2007). Mathematical models also revealed that certain delivery profiles can be useful for the optimization of the therapeutic index of chemotherapeutic drugs. Such profiles may include various factors such as the initial conditions of the host and tumor, the variability in circadian entrainment, as well as the length of cell cycle (Lévi et al. 2007).

Methods in systems biology are necessary for the elucidation of the chrono-interactions among various oscillating systems in order to provide accurate predictions of drug responses and optimal therapeutic strategies. Factors from multiple dimensions need to be included, such as the chronopharmacology networks and pathways and the synchronized cell cultures (Ortiz-Tudela et al. 2013). The variability among circadian entrainment and cell cycle is especially important, as the two systems interact at various levels from genes to proteins to pathways (Lévi et al. 2007; also see Chaps. 2–6). Specifically, systems biology models may help explain the concurrence between chrono-tolerance and chrono-efficacy to incorporate the variations in the circadian and cell cycle dynamics between the normal and disease cells. Such approaches would promote both tolerability and efficacy of the drugs and allow for personalized chronotherapy not only to inhibit the disease progression, but also to promote the quality of life (Ortiz-Tudela et al. 2013; Lévi et al. 2010).

7.2 Potential Circadian-Associated Biomarkers at Various Levels

7.2.1 Circadian-Associated Biomarkers at the Molecular and Cellular Levels

Because the circadian patterns are strongly associated with psychophysiological and pathological activities, the identification of biomarkers in circadian dysregulations can be applied in scientific, clinical, and translational studies. Circadian-associated biomarkers can provide dynamic and precise detections for both diagnosis and therapeutic outcome assessments.

At the molecular and cellular levels, genes regulating circadian rhythms are often involved in controlling other cellular pathways including cell cycles, cell proliferation, and apoptosis (Zhu et al. 2005). As circadian genes play critical roles in tumorigenesis and tumor growth via their involvement in cancer-associated pathways, they have been considered potential biomarkers for cancer risk detection, diagnosis, and prognosis (Yi et al. 2010). Table 7.1 provides examples of some potential circadian-associated biomarkers at the molecular and cellular levels for various conditions and diseases. A more complete list can be found in the Database of Biological Rhythms, including genomic, proteomic, as well as metabolomics biomarkers (DBBR 2015).

For example, polymorphisms in the circadian gene NPAS2 have been related to breast cancer development. A genotypic analysis of 348 breast cancer tissue samples showed that higher expression of NPAS2 was related to disease free survival and

Table 7.1 Examples of potential circadian-associated biomarkers at the molecular and cellular levels

Potential biomarkers	Associated conditions	Applications	References
ARNTL (also known as BMAL1)	Breast cancer, colorectal cancer	Prognosis, lymph node metastasis	Zeng et al. (2014) and Kuo et al. (2009)
BDNF	Stress-associated conditions	Personality traits, mood	Tirassa et al. (2012)
CLOCK	Breast cancer	Prognosis (survival)	Kuo et al. (2009)
CRY1	Colorectal cancer	Prognosis (lymph node metastasis, TNM stages, overall survival)	Yu et al. (2013))
CRY2	Colorectal cancer, non-Hodgkin's lymphoma (NHL)	Prognosis, risk predictions	Fang et al. (2015) and Hoffman et al. (2009)
NPAS2 (variants)	Breast cancer, colorectal cancer	Prognosis (TNM stages, tumor metastasis, tumor size, overall survival)	Xue et al. (2014) and Yi et al. (2010)
PER1	Non-small cell lung cancer (NSCLC)	Prognosis	Liu et al. (2014)
PER2	Breast cancer, gastric cancer, NSCLC	Prognosis, lymph node metastasis	Zhao et al. (2014), Liu et al. (2014), and Kuo et al. (2009)
PER3 (variants)	Breast cancer	Risk detections	Zhu et al. (2005)
RORA	Depression	Antidepressant treatment responses	Hennings et al. (2015)
TIMELESS	Breast cancer	Cancer risks, prognosis (tumor stages)	Mao et al. (2013)
NPSR1 pathways (MAPK, TGFβ)	Neuroendocrine tumors (NETs)	Diagnosis	Pulkkinen et al. (2014)
Serotonin-N-acetylserotonin-melatonin pathway	Autism spectrum disorders (ASDs)	Diagnosis	Pagan et al. (2014)

overall survival (Yi et al. 2010). In addition, various polymorphism genotypes were differentially correlated with the tumor severity. In another example, the analysis of 389 Caucasian breast cancer cases showed that a PER3 variant was related to higher risks for breast cancer among premenopausal women (Zhu et al. 2005). These findings indicate that the circadian genotypic profiles can become potential diagnostic and prognostic biomarkers for breast cancer.

In psychiatric problems, altered circadian genes may serve as potential biomarkers for bipolar disorders, including ARNTL1 (also known as BMAL1), CLOCK, NPAS2, NR1D1, and PER3 (Milhiet et al. 2011, 2014). Based on the analysis of disrupted circadian patterns, systemic biomarker profiles for bipolar susceptibility can be established using various parameters such as melatonin levels and the record of sleep/wake patterns (Milhiet et al. 2014).

In addition to the core clock genes, the circadian rhythmicity of other proteins such as cytokines may also be important biomarkers. For instance, the diurnal serum changes of inflammatory biomarkers are significant in patients with multiple sclerosis (MS) (Wipfler et al. 2013). Elevated serum levels of soluble vascular adhesion molecule-1 (sVCAM-1) around noon and in the early afternoon have been found in MS patients with active disease (Wipfler et al. 2013).

Furthermore, neuroendocrine factors such as melatonin and cortisol have essential roles in cellular metabolism. The daily profiles of abnormal circadian patterns of salivary melatonin and cortisol levels may be potential biomarkers associated with metabolic syndrome elements (Corbalán-Tutau et al. 2014). These metabolic syndrome elements include blood pressure, glucose and plasma lipids, ghrelin, as well as adipocyte-generated hormones including leptin and adiponectin. Such complex factors and elements may provide the connections across the molecular, cellular, and organism levels (see the next section).

7.2.2   Circadian-Associated Biomarkers at the Tissue and Organism Levels

Together with molecular and cellular biomarkers, circadian-associated factors at the tissue and organism levels may provide more complete views for systems-based studies. With multi-directional connections across various levels, these factors can be integrated in comprehensive profiles. For example in bipolar disorders, circadian deregulations are prominent properties during both acute mood episodes and euthymic phases as potential markers (Milhiet et al. 2011). Such circadian disruptions can be detected both quantitatively and qualitatively using various tools including polysomnography and blood melatonin monitoring. Other altered circadian parameters include the activities of fibroblasts and the core temperature (Milhiet et al. 2011).

In Parkinson disease (PD), patients with depression may have abnormal circadian patterns of the core body temperature as a potential biomarker (Suzuki et al. 2007). In addition, studies using rodent models showed that alterations in circadian profiles including the body temperature and locomotor activity may be biomarkers of the aging process (Mailloux et al. 1999).

Furthermore, the circadian rhythm "chaos" has been considered a prominent marker of breast cancer risks even before the mammographic proof of neoplasm or the detection of a palpable tumor (Keith et al. 2001). For instance, abnormal melatonin levels have been related to nightshift work and exposure to light-at-night in both experimental and clinical settings (Mirick and Davis 2008). It has been suggested as a robust biomarker of circadian disruptions under various environmental impacts. The direct and indirect measurements of the metabolites of melatonin in urine, blood, and saliva have been found reliable in epidemiologic research and studies of cancers with long inactive stages (Mirick and Davis 2008).

The circadian-associated biomarkers are also valuable in the assessment of therapeutic outcomes. In anticancer therapy, the circadian timing of medications may promote the therapeutic tolerability up to fivefold with double efficacy in both laboratory and clinical settings (Scully et al. 2011). Many factors may influence the effects of the chronotherapeutic schedule, including the molecular elements of the circadian timing system (CTS), gender, and drug dosages. Robust circadian biomarkers based on non-invasive methods such as the varied skin surface temperature rhythms can be used to monitor the CTS for dynamic assessments in personalized chronotherapy (Scully et al. 2011).

7.3 Clinical Applications of Chrono-Biomarkers and Personalized Chronotherapy

7.3.1 Chrono-Biomarkers and Chronotherapy in Cancer

At the molecular and cellular levels, the circadian clocks have impacts on nucleotide excision repair, DNA damage checkpoints, and apoptosis (Sancar et al. 2015; also see Chap. 6). The disruptions of the clocks may cause alterations in these processes and lead to cancer regression. Factors in these processes such as the activities of DNA repair enzymes may also influence the pharmacokinetics, pharmacodynamics, and the overall efficacy of anti-cancer drugs (Sancar et al. 2015).

Models of the circadian transcriptional loops can be useful for the personalization of cancer chronotherapy. For example, as a topoisomerase I inhibitor against colorectal cancer, irinotecan has circadian toxicity patterns (Li et al. 2013). The host-specific analysis of liver and colon expressions of NR1D1 and BMAL1 in the circadian transcriptional loops may be useful for detecting the chronotoxicity patterns to promote the drug tolerability (Li et al. 2013). In addition, P-glycoprotein (P-gp) has been considered a crucial factor in the circadian balance between toxicity and efficacy of irinotecan (Filipski et al. 2014). The inhibition of P-gp has been suggested applicable for the optimization of irinotecan chronotherapy.

As another example, the transcription factor hepatic leukemia factor (HLF) is involved in the regulation of circadian rhythms and cell death with abnormal expression in human cancers (Waters et al. 2013). A microarray analysis showed that HLF is associated with a complex transcriptional network involving the upregulation of anti-apoptotic genes together with the downregulation of pro-apoptotic genes (Waters et al. 2013). Because the altered expression of HLF has been correlated with both circadian patterns and higher resistance to cell death, it can be an important biomarker and target for chronotherapy.

In addition, endocrine therapeutic resistance is an important obstacle to the effective treatment of breast cancer. Melatonin has been found as a suppressor of tumor metabolism and a circadian-controlled kinase suppressor that may restore the sensitivity of breast tumors to tamoxifen (Dauchy et al. 2014). However, light exposure at night (LEN) may inhibit the nocturnal generation of melatonin and result in tumors insensitive to tamoxifen. Such interconnections among the circadian and melatonin disruptions, LEN, and tamoxifen resistance may be critical for improving the efficacies of cancer treatment.

Studies using both experimental models and cancer patients have shown that the circadian timing of drug administration may have remarkable effects on cancer chemotherapies (Ortiz-Tudela et al. 2013). Analyses of both single-agent and combination chemotherapies have indicated that the optimal antitumor efficacy often occur when the drugs are administered close to the periods with the best tolerability. The tolerability of the anticancer chemotherapies may fluctuate from two to tenfolds caused by circadian timing (Lévi et al. 2010). According to some international randomized clinical trials, significant improvement in tolerability can be achieved when cancer patients are given the same sinusoidal chronotherapy schedule over 24 h (Ortiz-Tudela et al. 2013). Such results were observed when the groups were compared to those who were given constant-rate infusions or incorrectly timed treatments.

7.3.2 Personalized Chronotherapy in Hypertension

The application of chronotherapy may provide benefits for the treatment of various diseases, especially hypertension. Chronotherapy has been deemed as a cost-effective strategy for improving blood pressure (BP) control during both nighttime sleep and daytime activities (Smolensky et al. 2010). For example, for patients with hypertension, bedtime dosing of the antihypertensive medications have been found to decrease cardiovascular morbidity (Schillaci et al. 2015). In resistant hypertension, the timing of treatment in accordance with the circadian BP pattern has been considered more important for the effectiveness than the simple adjustment of drug combinations (Hermida et al. 2008).

The treatment-time differences have been found clinically meaningful in various classes of medications. Specifically, calcium channel blockers may be more effective with bedtime than morning dosing. For instance, the bedtime dosing of dihydropyridine derivatives may remarkably decrease the risks of edema (Smolensky et al. 2010). Bedtime rather than morning dosing of angiotensin II receptor blockers and angiotensin-converting enzyme inhibitors may enhance the decline of the sleep-time BP and decrease the urinary albumin production with improved renal functions (Smolensky et al. 2010).

Better efficacies can also be achieved with evening dosing for various combinations of antihypertensive drugs. Such combinations include calcium channel blocker with angiotensin-receptor blocker, calcium channel blocker with diuretic, as well as angiotensin-receptor blocker with diuretic (Potúcek and Klimas 2013). In essential hypertension, chronotherapy of the valsartan/hydrochlorothiazide (HCTZ) combination and the valsartan/amlodipine combination given at bedtime may improve the sleep-time BP control (Hermida et al. 2010, 2011). For refractory arterial hypertension (RAH) associated with a non-dipping BP pattern and a poor clinical prognosis, adjusting the treatment time to evening has also been found more effective (Almirall et al. 2012).

Personalized chronotherapy is necessary for the optimization of the drug effects via different timing along the circadian scales. For example, some studies showed that bedtime administration of antihypertensive medications such as the renin-angiotensin-aldosterone system blockers may lower both cardiovascular and cerebrovascular risks (Hermida et al. 2013, 2014). However, in some other studies, the optimal treatment time varied significantly among different patients (Watanabe et al. 2013).

7.3.3   Chronotherapy in Other Diseases

As a rapidly developing area, chronotherapy has been applied in many diseases including psychiatric disorders. For instance, triple chronotherapy is a combination of total sleep deprivation (wake therapy), sleep phase advance, and bright light therapy (Sahlem et al. 2014). It has shown quick and continuous antidepressant effects (Gottlieb and Terman 2012). A recent study found that the adjunctive triple chronotherapy is practical and tolerable for quickly improving mood and reducing suicidality among acutely suicidal and depressed patients (Sahlem et al. 2014). Such adjunctive chronotherapy may also quicken and maintain antidepressant responses, supporting a safer and sustainable therapy for bipolar disorder (BPD) (Wu et al. 2009). In addition to depression, adjunctive chronotherapy may promote the effects of cognitive-behavioral therapy for patients with treatment-resistant obsessive compulsive disorder (OCD) (Coles and Sharkey 2011).

In rheumatoid arthritis (RA), circadian patterns have been associated with the occurrence and severity of chronic symptoms as well as morbid and mortal states (Haus et al. 2012). The severity of the symptoms of pain, stiffness and functional disability often reaches the peak in the early morning, which may be associated with the higher levels of proinflammatory cytokines especially IL6 (Spies et al. 2011). The practice of RA chronotherapy in the United States and Europe has a history of more than 50 years. Such practice included the administration of nonsteroidal anti-inflammatory drugs (NSAIDs), disease modifying antirheumatic drugs (DMARDs), and synthetic corticosteroid medications (Haus et al. 2012).

For example, low-dose prednisone chronotherapy using a modified-release (MR) formulation together with the DMARDs may lead to quick and significant reliefs in RA symptoms such as morning joint stiffness (Buttgereit et al. 2013; Spies et al. 2011). In this paradigm, prednisone is released about 4 h following ingestion, around 2 a.m. if the drug is taken at 10 p.m. in the evening (Spies et al. 2011). The MR prednisone administered in accordance with the biological rhythms may have beneficial effects on the HPA axis and the regulation of proinflammatory cytokines such as IL6 (Alten 2012).

Furthermore, prednisone chronotherapy using the MR tablet was found safe with higher tolerability, providing a continuous improvement with a better benefit-to-risk ratio for at least 12 months (Buttgereit et al. 2010). Studies using mouse models and RA patients showed that bedtime chronotherapy of Methotrexate (MTX) had better effects in improving RA symptoms and functional capacity than the standard dosing methods (To et al. 2011). The methods in such chronotherapy may reduce the levels of inflammation and plasma TNFα. With more studies in the relevant areas, more strategies in personalized chronotherapy can be developed to support the practice of personalized and systems medicine.

References

Almirall J, Comas L, Martínez-Ocaña JC, Roca S, Arnau A (2012) Effects of chronotherapy on blood pressure control in non-dipper patients with refractory hypertension. Nephrol Dial Transplant 27:1855–1859

Alten R (2012) Chronotherapy with modified-release prednisone in patients with rheumatoid arthritis. Expert Rev Clin Immunol 8:123–133

Buttgereit F, Doering G, Schaeffler A, Witte S, Sierakowski S, Gromnica-Ihle E, Jeka S, Krueger K, Szechinski J, Alten R (2010) Targeting pathophysiological rhythms: prednisone chronotherapy shows sustained efficacy in rheumatoid arthritis. Ann Rheum Dis 69: 1275–1280

Buttgereit F, Mehta D, Kirwan J, Szechinski J, Boers M, Alten RE, Supronik J, Szombati I, Romer U, Witte S et al (2013) Low-dose prednisone chronotherapy for rheumatoid arthritis: a randomised clinical trial (CAPRA-2). Ann Rheum Dis 72:204–210

Coles ME, Sharkey KM (2011) Compulsion or chronobiology? A case of severe obsessive-compulsive disorder treated with cognitive-behavioral therapy augmented with chronotherapy. J Clin Sleep Med 7:307–309

Corbalán-Tutau D, Madrid JA, Nicolás F, Garaulet M (2014) Daily profile in two circadian markers "melatonin and cortisol" and associations with metabolic syndrome components. Physiol Behav 123:231–235

Dallmann R, Brown SA, Gachon F (2014) Chronopharmacology: new insights and therapeutic implications. Annu Rev Pharmacol Toxicol 54:339–361

Dauchy RT, Xiang S, Mao L, Brimer S, Wren MA, Yuan L, Anbalagan M, Hauch A, Frasch T, Rowan BG et al (2014) Circadian and melatonin disruption by exposure to light at night drives intrinsic resistance to tamoxifen therapy in breast cancer. Cancer Res 74:4099–4110

DBBR (2015) The database of biological rhythms. http://pharmtao.com/health/biological-rhythms-database/. Accessed 1 June 2015

Fang L, Yang Z, Zhou J, Tung J-Y, Hsiao C-D, Wang L, Deng Y, Wang P, Wang J, Lee M-H (2015) Circadian clock gene CRY2 degradation is involved in chemoresistance of colorectal cancer. Mol Cancer Ther 14:1476–1487

Filipski E, Berland E, Ozturk N, Guettier C, van der Horst GTJ, Lévi F, Okyar A (2014) Optimization of irinotecan chronotherapy with P-glycoprotein inhibition. Toxicol Appl Pharmacol 274:471–479

Gottlieb JF, Terman M (2012) Outpatient triple chronotherapy for bipolar depression: case report. J Psychiatr Pract 18:373–380

Haus E, Sackett-Lundeen L, Smolensky MH (2012) Rheumatoid arthritis: circadian rhythms in disease activity, signs and symptoms, and rationale for chronotherapy with corticosteroids and other medications. Bull NYU Hosp Jt Dis 70(Suppl 1):3–10

Hennings JM, Uhr M, Klengel T, Weber P, Pütz B, Touma C, Czamara D, Ising M, Holsboer F, Lucae S (2015) RNA expression profiling in depressed patients suggests retinoid-related orphan receptor alpha as a biomarker for antidepressant response. Transl Psychiatry 5:e538

Hermida RC, Ayala DE, Fernández JR, Calvo C (2008) Chronotherapy improves blood pressure control and reverts the nondipper pattern in patients with resistant hypertension. Hypertension 51:69–76

Hermida RC, Ayala DE, Fontao MJ, Mojón A, Fernández JR (2010) Chronotherapy with valsartan/amlodipine fixed combination: improved blood pressure control of essential hypertension with bedtime dosing. Chronobiol Int 27:1287–1303

Hermida RC, Ayala DE, Mojón A, Fontao MJ, Fernández JR (2011) Chronotherapy with valsartan/hydrochlorothiazide combination in essential hypertension: improved sleep-time blood pressure control with bedtime dosing. Chronobiol Int 28:601–610

Hermida RC, Ayala DE, Smolensky MH, Mojón A, Fernández JR, Crespo JJ, Moyá A, Ríos MT, Portaluppi F (2013) Chronotherapy improves blood pressure control and reduces vascular risk in CKD. Nat Rev Nephrol 9:358–368

Hermida RC, Smolensky MH, Ayala DE, Fernández JR, Moyá A, Crespo JJ, Mojón A, Ríos MT, Fabbian F, Portaluppi F (2014) Abnormalities in chronic kidney disease of ambulatory blood pressure 24 h patterning and normalization by bedtime hypertension chronotherapy. Nephrol Dial Transplant 29:1160–1167

Hoffman AE, Zheng T, Stevens RG, Ba Y, Zhang Y, Leaderer D, Yi C, Holford TR, Zhu Y (2009) Clock-cancer connection in non-Hodgkin's lymphoma: a genetic association study and pathway analysis of the circadian gene cryptochrome 2. Cancer Res 69:3605–3613

Kaur G, Phillips C, Wong K, Saini B (2013) Timing is important in medication administration: a timely review of chronotherapy research. Int J Clin Pharm 35:344–358

Keith LG, Oleszczuk JJ, Laguens M (2001) Circadian rhythm chaos: a new breast cancer marker. Int J Fertil Womens Med 46:238–247

Kuo S-J, Chen S-T, Yeh K-T, Hou M-F, Chang Y-S, Hsu NC, Chang J-G (2009) Disturbance of circadian gene expression in breast cancer. Virchows Arch 454:467–474

Lévi F, Filipski E, Iurisci I, Li XM, Innominato P (2007) Cross-talks between circadian timing system and cell division cycle determine cancer biology and therapeutics. Cold Spring Harb Symp Quant Biol 72:465–475

Lévi F, Okyar A, Dulong S, Innominato PF, Clairambault J (2010) Circadian timing in cancer treatments. Annu Rev Pharmacol Toxicol 50:377–421

Li X-M, Mohammad-Djafari A, Dumitru M, Dulong S, Filipski E, Siffroi-Fernandez S, Mteyrek A, Scaglione F, Guettier C, Delaunay F et al (2013) A circadian clock transcription model for the personalization of cancer chronotherapy. Cancer Res 73:7176–7188

Liu B, Xu K, Jiang Y, Li X (2014) Aberrant expression of Per1, Per2 and Per3 and their prognostic relevance in non-small cell lung cancer. Int J Clin Exp Pathol 7:7863–7871

Mailloux A, Benstaali C, Bogdan A, Auzéby A, Touitou Y (1999) Body temperature and locomotor activity as marker rhythms of aging of the circadian system in rodents. Exp Gerontol 34:733–740

Mao Y, Fu A, Leaderer D, Zheng T, Chen K, Zhu Y (2013) Potential cancer-related role of circadian gene TIMELESS suggested by expression profiling and in vitro analyses. BMC Cancer 13:498

Milhiet V, Etain B, Boudebesse C, Bellivier F (2011) Circadian biomarkers, circadian genes and bipolar disorders. J Physiol Paris 105:183–189

Milhiet V, Boudebesse C, Bellivier F, Drouot X, Henry C, Leboyer M, Etain B (2014) Circadian abnormalities as markers of susceptibility in bipolar disorders. Front Biosci (Schol Ed) 6:120–137

Mirick DK, Davis S (2008) Melatonin as a biomarker of circadian dysregulation. Cancer Epidemiol Biomarkers Prev 17:3306–3313

Ortiz-Tudela E, Mteyrek A, Ballesta A, Innominato PF, Lévi F (2013) Cancer chronotherapeutics: experimental, theoretical, and clinical aspects. Handb Exp Pharmacol 261–288

Pagan C, Delorme R, Callebert J, Goubran-Botros H, Amsellem F, Drouot X, Boudebesse C, Le Dudal K, Ngo-Nguyen N, Laouamri H et al (2014) The serotonin-N-acetylserotonin-melatonin pathway as a biomarker for autism spectrum disorders. Transl Psychiatry 4:e479

Potúcek P, Klimas J (2013) Chronotherapy of hypertension with combination treatment. Pharmazie 68:921–925

Pulkkinen V, Ezer S, Sundman L, Hagström J, Remes S, Söderhäll C, Greco D, Dario G, Haglund C, Kere J et al (2014) Neuropeptide S receptor 1 (NPSR1) activates cancer-related pathways and is widely expressed in neuroendocrine tumors. Virchows Arch 465:173–183

Sahlem GL, Kalivas B, Fox JB, Lamb K, Roper A, Williams EN, Williams NR, Korte JE, Zuschlag ZD, El Sabbagh S et al (2014) Adjunctive triple chronotherapy (combined total sleep deprivation, sleep phase advance, and bright light therapy) rapidly improves mood and suicidality in suicidal depressed inpatients: an open label pilot study. J Psychiatr Res 59:101–107

Sancar A, Lindsey-Boltz LA, Gaddameedhi S, Selby CP, Ye R, Chiou Y-Y, Kemp MG, Hu J, Lee JH, Ozturk N (2015) Circadian clock, cancer, and chemotherapy. Biochemistry 54:110–123

Schillaci G, Battista F, Settimi L, Schillaci L, Pucci G (2015) Antihypertensive drug treatment and circadian blood pressure rhythm: a review of the role of chronotherapy in hypertension. Curr Pharm Des 21:756–772

Scully CG, Karaboué A, Liu W-M, Meyer J, Innominato PF, Chon KH, Gorbach AM, Lévi F (2011) Skin surface temperature rhythms as potential circadian biomarkers for personalized chronotherapeutics in cancer patients. Interface Focus 1:48–60

Smolensky MH, Hermida RC, Ayala DE, Tiseo R, Portaluppi F (2010) Administration-time-dependent effects of blood pressure-lowering medications: basis for the chronotherapy of hypertension. Blood Press Monit 15:173–180

Spies CM, Cutolo M, Straub RH, Burmester GR, Buttgereit F (2011) Prednisone chronotherapy. Clin Exp Rheumatol 29:S42–S45

Suzuki K, Miyamoto T, Miyamoto M, Kaji Y, Takekawa H, Hirata K (2007) Circadian variation of core body temperature in Parkinson disease patients with depression: a potential biological marker for depression in Parkinson disease. Neuropsychobiology 56:172–179

Tirassa P, Iannitelli A, Sornelli F, Cirulli F, Mazza M, Calza A, Alleva E, Branchi I, Aloe L, Bersani G et al (2012) Daily serum and salivary BDNF levels correlate with morning-evening personality type in women and are affected by light therapy. Riv Psichiatr 47:527–534

To H, Yoshimatsu H, Tomonari M, Ida H, Tsurumoto T, Tsuji Y, Sonemoto E, Shimasaki N, Koyanagi S, Sasaki H et al (2011) Methotrexate chronotherapy is effective against rheumatoid arthritis. Chronobiol Int 28:267–274

Watanabe Y, Halberg F, Otsuka K, Cornelissen G (2013) Toward a personalized chronotherapy of high blood pressure and a circadian overswing. Clin Exp Hypertens 35:257–266

Waters KM, Sontag RL, Weber TJ (2013) Hepatic leukemia factor promotes resistance to cell death: implications for therapeutics and chronotherapy. Toxicol Appl Pharmacol 268:141–148

Wipfler P, Heikkinen A, Harrer A, Pilz G, Kunz A, Golaszewski SM, Reuss R, Oschmann P, Kraus J (2013) Circadian rhythmicity of inflammatory serum parameters: a neglected issue in the search of biomarkers in multiple sclerosis. J Neurol 260:221–227

Wu JC, Kelsoe JR, Schachat C, Bunney BG, DeModena A, Golshan S, Gillin JC, Potkin SG, Bunney WE (2009) Rapid and sustained antidepressant response with sleep deprivation and chronotherapy in bipolar disorder. Biol Psychiatry 66:298–301

Xue X, Liu F, Han Y, Li P, Yuan B, Wang X, Chen Y, Kuang Y, Zhi Q, Zhao H (2014) Silencing NPAS2 promotes cell growth and invasion in DLD-1 cells and correlated with poor prognosis of colorectal cancer. Biochem Biophys Res Commun 450:1058–1062

Yi C, Mu L, de la Longrais IAR, Sochirca O, Arisio R, Yu H, Hoffman AE, Zhu Y, Katsaro D (2010) The circadian gene NPAS2 is a novel prognostic biomarker for breast cancer. Breast Cancer Res Treat 120:663–669

Yu H, Meng X, Wu J, Pan C, Ying X, Zhou Y, Liu R, Huang W (2013) Cryptochrome 1 overexpression correlates with tumor progression and poor prognosis in patients with colorectal cancer. PLoS One 8:e61679

Zeng Z, Luo H, Yang J, Wu W, Chen D, Huang P, Xu R (2014) Overexpression of the circadian clock gene Bmal1 increases sensitivity to oxaliplatin in colorectal cancer. Clin Cancer Res 20:1042–1052

Zhao H, Zeng Z-L, Yang J, Jin Y, Qiu M-Z, Hu X-Y, Han J, Liu K-Y, Liao J-W, Xu R-H et al (2014) Prognostic relevance of Period1 (Per1) and Period2 (Per2) expression in human gastric cancer. Int J Clin Exp Pathol 7:619–630

Zhu Y, Brown HN, Zhang Y, Stevens RG, Zheng T (2005) Period3 structural variation: a circadian biomarker associated with breast cancer in young women. Cancer Epidemiol Biomarkers Prev 14:268–270

Index

© Springer International Publishing Switzerland 2015   83
Q. Yan, *Cellular Rhythms and Networks: Implications for Systems Medicine*,
SpringerBriefs in Cell Biology, DOI 10.1007/978-3-319-22819-8